MONOGRAPHS ON
APPLIED PROBABILITY AND STATISTICS

*General Editors*

M.S. BARTLETT, F.R.S., *and* D. COX, F.R.S.

# MULTIVARIATE ANALYSIS IN BEHAVIOURAL RESEARCH

# Multivariate Analysis in Behavioural Research

A. E. MAXWELL

*Professor of Statistics*
*Institute of Psychiatry, London*

LONDON
CHAPMAN AND HALL

A Halsted Press Book
John Wiley & Sons, New York

*First published* 1977
*by Chapman and Hall Ltd.*
11 *New Fetter Lane, London EC4P 4EE*



*Photosetting by Thomson Press (India) Limited, New Delhi*
*and printed in Great Britain by*
*Richard Clay (The Chaucer Press) Ltd.,*
*Bungay, Suffolk*

ISBN 0 412 14300 3



*Distributed in the U.S.A. by Halsted Press,*
*a Division of John Wiley & Sons, Inc., New York*

**Library of Congress Cataloging in Publication Data**

Maxwell, Albert Ernest.
Multivariate analysis in behavioural research.

(Monographs on applied probability and statistics)
Bibliography: p.
Includes index.
1. Psychometrics. 2. Multivariate analysis.
I. Title.
BF39.M36 150'.1'82 76-25110
ISBN 0-470-98902-5

# Contents

Preface *page* ix

**1. Historical Background** **1**
*Introduction* 1
*Principal component analysis* 2
*Factor analysis* 5
*Hierarchical correlation matrices* 8
*Additional factors* 10
*Naming factors* 10
*Explaining the low rank of correlation matrices* 11

**2. General Observations** **13**
*Introduction* 13
*Estimating variances and covariances* 13
*Linear constraints on variates* 16
*Matrices of reduced rank* 17
*Terminology* 18
*Metric* 19
*The distribution of variates* 20

**3. Matrices and Determinants** **22**
*Matrices and their manipulation* 22
*Determinants* 25
*Inverse matrices* 26
*Some uses of matrices and determinants* 27
*Rank of matrix* 29
*Latent roots and vectors of a matrix* 30
*Latent vectors* 31
*Orthogonal matrices* 33
*Matrix rotation* 34
*Triangular matrices* 34
*Quadratic forms and their differentiation* 35
*Lagrange multiplier* 36
*Latent roots and vectors of non-symmetric matrices* 37

**4. Principal Component Analysis** **39**
*Matrix transformation* 39

*Appraisal of results* 41
*Practical applications* 41
*Q-technique* 44
*Extracting a 'difficulty' component* 45

**5. Factor Analysis** **46**
*Review* 46
*The basic model* 46
*Using the model* 49
*The residual variates* 50
*Factor rotation* 53
*A salutary point* 57
*Factor analysis or principal component analysis?* 58

**6. Confirmatory Factor Analysis** **60**
*The restricted factor model* 60
Estimating Factor Scores 66
*Oblique factors* 68

**7. Multiple Linear Regression** **70**
*Introduction* 70
*The model* 71
*Scaling* 73
*Precautions* 74
*Selective sampling* 74
*Selecting independent variates* 78
*Appraisal* 79
*Regression and factor analysis* 81
*Addendum* 82

**8. Canonical Correlations** **85**
*Introduction* 85
*The model* 85
*A computational problem* 88
*Tests of significance* 88
*Inter-rater agreement* 91
*General comments* 93

**9. Discriminant Function and Canonical Variate Analysis** **94**
*Introduction* 94
*The case of two groups* 95
*Canonical variate analysis* 96
*Interpretation* 97
*Scaling of scores on the observed variates* 98
*Scaling the canonical variates* 99
*Quadratic discriminant functions* 102
*Comparing covariance matrices* 105

**10. The Analysis of Contingency Tables** **106**
*Introduction* 106
*Notation* 106
*Categories with a natural order* 111
*General comments* 113

**11. Analysis of Variance in Matrix Notation** **115**
*Introduction* 115
*Experimental designs* 115
*Alternative procedures* 119
*Non-orthogonal designs* 123
*Generalization* 126

**12. Multivariate Analysis of Variance (MANOVA)** **129**
*Introduction* 129
*The model* 129
*Partitioning the matrix* $\mathbf{\Theta}'\mathbf{D}\mathbf{\Theta}$ 132
*Tests of significance in MANOVA* 133

**13. Cluster Analysis and Miscellaneous Techniques** *by* B.S. Everitt **136**
*Introduction* 136
*Cluster analysis techniques* 136
*Similarity and distance measures* 137
*Common techniques* 137
*Evaluating solutions* 140
Visual Representations of Multivariate Data 143
*Summary* 150

**References** **153**

**Index** **159**

# Preface

This book is based on a course of lectures on multivariate analysis given by the author annually to postgraduate students and research workers in the behavioural sciences. It is introductory in two senses, firstly to prepare them for further instruction in the use of computer packages for the analysis of multivariate data and, secondly, to assist them in their study of more advanced books on the subject, such as that by Morrison. However, although introductory, the book is not elementary and the student who wishes to benefit from it must first acquire a working knowledge of matrix algebra by a careful reading of Chapter 3.

The bulk of the book is concerned with the classic techniques of multivariate analysis, but throughout, special attention in directed to the disruptive role which errors of measurement in observations can play when these techniques are employed, and to possible ways of assessing the effects of such errors. Chapter 11, on analysis of variance, is concerned primarily with univariate analysis but it includes many of the techniques required for multivariate analysis of variance which is discussed briefly in Chapter 12. Both chapters should serve as a helpful introduction to specialist books on this topic, such as those by Bock and Finn. The book ends with a chapter, by B.S. Everitt, on cluster analysis and other exploratory techniques which in recent years have become popular.

In the historical introduction exact references to work by Galton and other early writers are omitted as they refer to journals which are not readily accessible, but most of them can be found in the papers by Karl Pearson referred to in the text.

The author wishes to thank his colleagues at the Institute, especially Mr Brian Everitt and Mr Owen White, who have been of constant assistance during the writing of the book, and Mrs Bertha Lakey who typed the manuscript with great care and attention to detail.

*Institute of Psychiatry*
*University of London*
1975

A.E. MAXWELL

CHAPTER ONE

# Historical Background

## Introduction

As long ago as 1883, Francis Galton wrote that 'the object of statistical science is to discover methods of condensing large groups of allied facts into brief and compendious expressions suitable for discussion'. His statement is peculiarly apt in the case of multivariate analyses. Here the amount of data, which depends not only on the size of the sample but also on the number of variates being considered simultaneously, may be daunting. Yet multivariate statistical techniques seek to describe them in terms of a relatively small number of parameters and to facilitate their interpretation, often with considerable enlightenment, in terms of but a few derived variates. In introducing these techniques it is instructive to review briefly the early history of their development.

The origins of multivariate analysis are closely connected with the development of methods of correlation. While the idea of 'correlation' can be traced back to Aristotle, the concept as now employed in statistical work, is due to Galton himself: he was the first person to make it clear that a measure of concomitant variation between two measurable organs, traits or characteristics need not be perfect and was also the first to provide a procedure for measuring numerically the degree of the association between them. Karl Pearson (1920), in his paper *Notes on the history of correlation*, refers to a meeting of the Royal Institution of Great Britain held on Friday, 9th February, 1877, at which Galton gave a lecture under the title *Typical laws of heredity in man.* Concerning it Pearson says, "Here for the first time appears a numerical measure $r$ of what is termed 'reversion' which Galton later termed 'regression'. This $r$ is the source of our symbol for the correlation coefficient".

From this statement it is clear that the $r$ in question is what is now called the regression coefficient of one variable on another. But if both variables are measured in comparable units, say in units of the standard deviations of each, then the regression coefficient is numerically equal to the correlation coefficient, and Galton had indeed measured his variates in comparable units using their probable errors (or quantile values) as units of measurement.

In a later contribution Galton considered the more general problem of how to measure the degree to which one variate may be co-related with the combined effect of a number of further variates whether the latter be co-related or not. This is the problem of multiple correlation. Galton himself never succeeded to solving it satisfactorily, but it was soon taken up by Karl Pearson and Yule who, at the turn of the century, provided statistical procedures which are still in use to-day. These included the ideas of partial correlation and partial regression coefficients leading up to the multiple correlation coefficient itself. As it turned out this multiple or maximum coefficient was obtained by taking a weighted sum of the variates to be combined, a result which Galton had anticipated.

In the latter aspect of the early work on correlational analysis, it is important to emphasize that what was sought was the best set of weights to apply to a given set of variates to maximize the correlation of their combined effect with some other variate, or 'external criterion'. This was a clear-cut and well defined problem: but another problem, in many respects more difficult to appraise, was then exercising the minds of biometricians.

## Principal component analysis

The problem, in part at least, arose from an attempt to identify criminals by means of a set of body-measurements made on them. At the time the Home Office in London had provisionally adopted a measurement procedure suggested by the Frenchman, Bertillion, which involved twelve body measurements. Galton criticized the procedure on the grounds that several of the measurements, for example arm-length and leg-length, were so highly correlated as to render it uneconomical to measure both. What was needed he claimed were measures which correlated minimally with each other and so represented relatively independent indices for classification purposes. Galton's suggestion was taken up by Edgeworth who decided to push it to the limit and replace the observed measurements by a set of hypothetical variates derived from them which would have the property that they were uncorrelated. To illustrate the procedure he took three body measurements, namely *stature*, *forearm* and *leg-length*, for which correlations had already been worked out by Galton, and derived three hypothetical and linear functions of the variables, namely,

$$f_1 = 0.16 \text{ stature} + 0.51 \text{ forearm} + 0.39 \text{ leg-length},$$
$$f_2 = -0.17 \text{ stature} + 0.69 \text{ forearm} - 0.09 \text{ leg-length},$$
$$f_3 = -0.15 \text{ stature} - 0.25 \text{ forearm} + 0.52 \text{ leg-length},$$

which were approximately uncorrelated. Here then was a set of weighted functions, each resembling in a formal way the weighted function derived in a multiple regression problem, but obtained without reference to any outside criterion. It was a purely 'internal' analysis, which led to as many hypothetical variates as there were observed variates and it represents the first attempt to do a *component* analysis.

Meanwhile, one of Pearson's collaborators, W.R. Macdonell, had obtained from the Metric office, New Scotland Yard, measurements on seven physical variables for each of 3000 criminals. The results were analysed and reported, in a paper published in 1902. Part of the analysis consisted in estimating the correlations between the variables in pairs and these were printed in matrix form. As this was probably the very first time that a set of intercorrelations was presented in this way it seems worth reporting them here. The seven variables measured were

(1) Head length, (2) Head breadth,
(3) Face breadth, (4) Left finger length,
(5) Left cubit (forearm), (6) Left foot,
(7) Height.

The correlations were originally given to five decimal places but are reproduced here to three places only.

Table 1.1 Macdonell's correlation matrix

| *Variables* | (1) | (2) | (3) | (4) | (5) | (6) | (7) |
|---|---|---|---|---|---|---|---|
| (1) | 1.000 | 0.402 | 0.395 | 0.301 | 0.305 | 0.339 | 0.340 |
| (2) | 0.402 | 1.000 | 0.618 | 0.150 | 0.135 | 0.206 | 0.183 |
| (3) | 0.395 | 0.618 | 1.000 | 0.321 | 0.289 | 0.363 | 0.345 |
| (4) | 0.301 | 0.150 | 0.321 | 1.000 | 0.846 | 0.759 | 0.661 |
| (5) | 0.305 | 0.135 | 0.289 | 0.846 | 1.000 | 0.797 | 0.800 |
| (6) | 0.339 | 0.206 | 0.363 | 0.759 | 0.797 | 1.000 | 0.736 |
| (7) | 0.340 | 0.183 | 0.345 | 0.661 | 0.800 | 0.736 | 1.000 |

The problem Macdonell wished to solve was how the data could most efficiently be used for purposes of identifying a criminal. He comments that 'the conditions to be satisfied in selecting a series of organs for criminal identification are:

(i) comparative ease and accuracy of measurement,

(ii) small correlation between them, so that fairly few organs will provide a reliable index to the criminal population'. He remarks that the organs selected by Scotland Yard are 'very far from being slightly correlated' and that the seven variates in question would hardly have been chosen if this information had been known in advance. He proceeded to set up regression equations for predicting any particular variate from some or all of the remaining six, working on the assumption that 'the best organ to leave to the last will be that, the variability of which is least reduced by selecting the other six.'

The analysis of Macdonell's data was carried out at the Galton Laboratory and it is clear from his paper that at the time Pearson was already convinced that the ideal indices to use would be those corresponding to the perpendicular axes of the multidimensional ellipsoid obtained when measurements are plotted in $p$-dimensional space. He says, 'Professor Pearson has pointed out to me that the ideal index characters would be given if we calculated the seven directions of uncorrelated variables, that is, the principal axes of the correlation *ellipsoid*.'

Pearson's procedure is described in his 1901 paper entitled, *On lines and planes of closest fit in systems of points in space*, in which a theoretical solution to the problem of principal axes, in terms of least squares, is given. Pearson illustrates the procedure by two simple examples but makes no suggestions as to how the routine calculations might be performed when the number of variates was relatively large. For a more practical discussion of the principal axes problem and for routine methods of calculation we have to turn to Hotelling's well-known papers on the *Analysis of a complex of statistical variables into principal components*, and the references therein (Hotelling, 1933).

As we have now seen the basis for the simultaneous analysis of several variables was already laid at the beginning of this century. One hesitates to say that the basis for multivariate analysis was thus laid for statisticians to-day prefer to date this from the work of later writers, namely J. Wishart who first gave us the distribution of the generalized product moment function in samples from normal multivariate populations; H. Hotelling who provided the generalized *t*-distribution, and S.S. Wilks and M.S. Bartlett who provided additional tests of hypotheses on means, variance and covariances. The latter papers appeared some thirty to forty years after the early contributions by Galton and Pearson and were inspired by the pioneering work of R.A. Fisher on small sample theory. It was in this intervening period that factor analysis, as distinct from component analysis, came on the scene.

**Factor analysis**

While the basis of factor analysis, in common with that of principal component analysis, undoubtedly lies in the early work by Galton and Pearson on correlation analysis, the motivation for its development was somewhat different and is to be found in the theories then current about the nature of 'mental power and faculties'. Drawing an analogy between mental ability and the distribution of physical characteristics, such as height and chest measurement in human populations, which he had shown to his satisfaction followed 'the law of deviation from an average' (the normal distribution law), Galton (1869) in his book *Hereditary Genius* asserts that the 'analogy clearly shows there must be a fairly constant average mental capacity in the inhabitants of the British Isles, and that the deviations from that average – upwards towards genius, and downwards towards stupidity – must follow the law that governs deviations from all true averages'. This assertion, which was later taken up and considerably elaborated by Pearson (1904), implies that there is a continuum along which 'mental capacity' can be measured and that in the population it is approximately normally distributed. The term universally used today for the variable implied is 'intelligence'.

Now the concept of intelligence has been, and indeed still is, a controversial one. In a succinct and scholarly article Burt (1955) traces its growth from classical writers including Plato, Aristotle and Cicero down to Herbert Spencer, and then turns to present-day biological, physiological and statistical evidence which testifies to the validity of the concept. As a result of his survey he concludes, 'that there is a general factor making for efficiency in all mental activities, and this factor is essentially cognitive or directive, and that the greater part of the individual variance found in this factor is attributable to differences in genetic constitution'. But he reminds us that 'it is essential to distinguish between intelligence as an abstract component of the individuals' genetic constitution and intelligence as an empirically measurable trait'. Factor analysis had its origin in an attempt to define this trait objectively and to measure it. The pioneer in this field was Charles Spearman.

### *Spearman's two-factor theory*

Following Spencer and Galton, Spearman appears to have adopted the view that the ability to discriminate was the essence of intelligence. He notes that the tests which Galton himself favoured were tests of sensory discrimination which could be administered under controlled conditions and so would be likely to furnish more or less precise objective information. But he also felt that measurements obtained in this way were far removed from what was generally understood by intelligence. To reconcile the two ideas the test results had to be checked against criteria such as teachers' estimates of the relative abilities of their students, and against marks allotted to the students for school subjects

Spearman was much impressed by the correlational techniques put forward by Galton and Pearson, but he had reservations about their uncritical use in psychological investigations. No matter how carefully we try to assess traits such as intelligence or the ability to discriminate our assessments, he emphasized, will never have the precision or accuracy (reliability) that attended the measurement of physical characteristics such as weight or height. To help remedy this defect he recommended obtaining for each laboratory test (or other psychological variable) a measure of its reliability by correlating the results obtained from a sample of subjects on two separate administrations of the test. Given this information, he showed (Spearman, 1904a) how an observed correlation between two different traits could, on average, be corrected for attenuation due to errors in their measurement.

But the correction of an observed correlation for errors of measurement was, in Spearman's mind, only one of the refinements which should be undertaken. He complains that in much of the experimental work he has surveyed, the authors have tried to kill as many birds as possible with one stone and have sought after the greatest – instead of the least – diversity: they have purposely thrown together subjects of all sorts and ages, and thus have gone out of their way to invite fallacious elements into their work. He thus decided in his own work to eliminate, by the use of partial correlation techniques and other means, sources of contamination due to variation in age between subjects, differences due to sex, and other extraneous influences.

In a second paper in 1904 Spearman reports the results of several investigations which he himself had carried out. One of these is based on 23 boys from a preparatory school. They had been administered a test of pitch discrimination, and scores or ratings for them on several school subjects – Classics, French, English, Mathematics and Music – were also obtained. The inter-correlations of the scores were then calculated and after corrections for variations in age and errors of measurement the correlations are reported as follows:

Table 1.2 Correlation matrix due to Spearman

| | *Classics* | *French* | *English* | *Math.* | *Discrim.* | *Music* |
|---|---|---|---|---|---|---|
| *Classics* | – | 0.83 | 0.78 | 0.70 | 0.66 | 0.63 |
| *French* | 0·83 | – | 0.67 | 0.67 | 0.65 | 0.57 |
| *English* | 0.78 | 0.67 | – | 0.64 | 0.54 | 0.51 |
| *Math.* | 0.70 | 0.67 | 0.64 | – | 0.45 | 0.51 |
| *Discrim.* | 0.66 | 0.65 | 0.54 | 0.45 | – | 0.40 |
| *Music* | 0.63 | 0.57 | 0.51 | 0.51 | 0.40 | – |

In this table the variates have been arranged so that the one which correlates on average most highly with the others is placed first, and so on. On examining the table thus arranged it is seen that the correlations show an interesting pattern: for any selected coefficient all those to its right in the same row, or below it in the same column (with minor exceptions) progressively decrease. To quote Spearman's own words, 'if we consider the correspondence between the four branches of school study a very remarkable uniformity may be observed... The whole forms *a perfectly constant Hierarchy* in the following order: Classics, French, English and Mathematics. This unbroken regularity becomes especially astonishing when we regard the minuteness of the variations involved, for the four branches have average correlations of 0.77, 0.72, 0.70 and 0.67 respectively. When in the same experimental series we turn to the Discrimination of Pitch, we find its correlations to be of slightly less magnitude (raw) but in precisely the same relative rank; being:

0.66 with Classics, 0.65 with French, 0.54 with English, 0.45 with Mathematics. The same is true of Musical Talent, a factor which is usually set up on a pedestal entirely apart'.

On the basis of these and other data, Spearman became convinced that, provided the correlations between the variables used were purified of contaminating influences, a close correspondence did in fact exist between 'all forms of sensory discrimination and the more complicated intellectual activities of practical life'. This correspondence he claimed was independent of the types of cognitive measures used and was 'reproducible in all times, places and manners'.

'All branches of intellectual activity', he asserted, 'have in common one fundamental function (or group of functions), whereas the remaining or specific elements of the activity seem in every case to be wholly different from that in all the others'. 'As an important practical consequence of this universal Unity of Intellectual function, the various actual forms of mental activity constitute a stably interconnected Hierarchy according to their different degrees of intellectual saturation.'

**Hierarchical correlation matrices**

Spearman's paper had a very stimulating effect on the psychological world and over the next few decades the full implications of his assertions were threshed out. Put briefly, the hierarchical structure which he had observed in his matrices of correlations, if indeed it were true, could most readily be explained by postulating a single hypothetical mental faculty entering into the performance of each school subject or laboratory test of discriminative ability. Many years earlier Galton had pointed out that if the sole 'cause' of correlation between two traits $i$ and $j$ was a third trait $g$, then the relationship

$$r_{ij} = r_{ig} r_{jg} \tag{1.1}$$

must hold. Galton's contention was later shown to be valid when the formula for a partial correlation coefficient

$$r_{ij.g} = \frac{r_{ij} - r_{ig} r_{jg}}{\sqrt{(1 - r_{ig}^2)}\sqrt{(1 - r_{jg}^2)}} \tag{1.2}$$

was worked out, for if $g$ were the sole cause of the correlation between $i$ and $j$, and if $g$ were held constant then $r_{ij,g}$ would become zero and formula (1.1) would immediately follow.

Equation (1.1) is the basic equation in Spearman's two-factor theory. Expressed in words it says that the correlation between two tests is given by the product of their separate correlations with the hypothetical factor $g$. If this holds true for all the correlations in a matrix then another interesting equation can be set up. As a simple illustration consider the correlations between tests (1) and (2), with (3) and (4). They are

| | (3) | (4) |
|---|---|---|
| (1) | $r_{13}$ | $r_{14}$ |
| (2) | $r_{23}$ | $r_{24}$ |

By the use of equation (1.1) it is easy to show that

$$r_{13}r_{24} - r_{23}r_{14} = 0.$$

The left hand side of equation (1.3) is known as the tetrad difference of the four correlations, and on the assumption that a single common factor underlies the tests, it has the value zero. For generality we may call the tests $i$ and $j$ and $s$ and $t$ and the tetrad difference then becomes

$$r_{is}r_{jt} - r_{js}r_{it}. \tag{1.3}$$

This expression is known mathematically as the value of a minor determinant of order two of the matrix, and when all such minor determinants are zero the matrix is said to be of rank *one*.

If in equation (1.1) we set $j = i$ we get

$$r_{ii} = r_{ig}^2. \tag{1.4}$$

The term $r_{ii}$ came to be known as the *self-correlation* or *communality* of test $i$, and $r_{ig}^2$ represented the amount of the variance of the test which could be accounted for by the hypothetical factor $g$, common to all the tests of the battery. Since the diagonal cells in a correlation matrix are unity the discrepancy $(1 - r_{ig}^2)$ was called the *specific* variance of the test. In other words the variance of each test, after eliminating error, was taken to be composed of two independent parts, one due to the common factor, the other due to a factor specific to the particular test itself.

## Additional factors

In the years following 1904 several investigations, along the lines indicated by Spearman, were carried out. Five years later Burt (1909) published the results of an extensive study entitled *Experimental Tests of General Intelligence*, one of the aims of which was to see if "higher mental functions would not show a yet closer connection with 'General Intelligence' than was shown by simpler mental functions, such as sensory discrimination and motor reaction, with which previous investigators had been so largely engrossed". Overall, Burt found strong evidence of hierarchical structure in his correlation matrix, though there was suggestive evidence of a sensory discrimination factor over and above the general factor. The latter finding was confirmed by the work of several American writers.

Bit by bit Spearman's theory came to be modified. In 1917, Burt provided clear evidence of verbal, numerical and practical group factors in school subjects in addition to a general factor. Kelley (1928) provided evidence to support this with a list of *verbal*, *number*, *rote memory*, *spatial* and *speed* factors. The numerous other studies which cast doubt on Spearman's claim need not be mentioned in detail but evidence of a complete break, with the two-factor theory, came with the publication by Thurstone (1938) of the first of a series of studies in which the notion of a general factor – though it existed – was ignored completely. Thurstone preferred to incorporate it in a series of overlapping group factors, and his approach set the pattern for most of the subsequent work by American writers. On the other hand many British writers, in particular Burt and Vernon and their students, tended to retain the concept of a general factor which, they maintained, is cognitive, general and innate, but to supplement it with further group (or overlapping group) factors sufficient to account adequately for the *covariation* of the variates being analysed.

## Naming factors

Although Spearman was loath to say precisely what the nature of the general factor postulated was he did suggest that it could be thought of as 'mental energy'. But later writers, as we have seen, were less hesitant about 'naming' factors referring to them by such labels as *verbal*, *rote memory*, and so on – labels which were generic in the sense that they stood for the class of test which had relatively high loadings on the factor concerned. Indeed Thurstone (1938) made so bold as to refer to the factors he had isolated as 'primary mental abilities', they included *verbal comprehension, word fluence, number, space, associative memory, perceptual speed* and *reasoning,* names which reflect the 'faculties' of the earlier psychologists.

Unfortunately as new tests were added to a battery, or as new batteries of tests were developed, it became necessary to postulate sub-classes of the factors already identified and to suggest new names for additional factors which did not fit into existing categories. Indeed the quest for new factors now seems well-nigh endless as the work of Guilford (1967) and his co-workers shows.

By contrast with Thurstone's proposal to account for cognitive behaviour in terms of 'primary mental abilities', Burt maintained that the structure of the mind is hierarchical (though not in Spearman's sense). In his own words (Burt, 1949) – "the processes of the lowest level are assumed to consist of simple sensations or simple movements such as can be artificially isolated and measured by tests of sensory 'thresholds' and by the timing of 'simple reactions'. The next level includes the more complex processes of perception and co-ordinated movement as in experiments on the apprehension of form and pattern or on 'compound reactions'. The third is the associative level – the level of memory and of habit formation. The fourth and highest of all involves the apprehension or application of relations. 'Intelligence', as the 'integrative capacity of the mind', is manifested at every level, but these manifestations differ not only in degree, but also (as introspection suggests) in their qualitative nature."

It is not easy to assemble a battery of tests to illustrate Burt's scheme; perhaps the most successful attempt to do so is that by Moursy (1952). See also a reanalysis of Moursy's data by Maxwell (1972a).

## Explaining the low rank of correlation matrices

One of the most persistent critics of factor analysis, and one who at the same time had a profound understanding of the subject, was Godfrey Thomson. Repeatedly in his writings he warned psychologists about the dangers of reifying factors and of depicting them as *causal* entities in the minds of men, and he was dubious about the advisability of adopting a model which necessitated the idea of specific factors. But he was impressed by the fact that matrices of correlation coefficients between cognitive variables, after appropriate adjustment of the diagonal elements, did tend to have ranks lower than their order, and one of his most notable contributions was to suggest a possible model for the brain which would account for the phenomenon. His suggestion (Thomson, 1919, 1939) was that the mind might be regarded as the synthesis of a large number of small components or 'bonds'. He did not try to define the term 'bond' rigorously, suggesting only that it might be thought of as a 'connection' established when a habit is formed – an idea possibly taken over from Thorndike. Whatever the basic mental 'units' (bonds) might be, he suggested that in the performance of a test a random sample of them was called into play – a larger sample for a difficult test a smaller sample for a simpler test. On the basis of this model he was able to show that if in a battery of $n$ tests the proportions of the total number of the mind's bonds being sampled were, $p_1, p_2, \ldots, p_n$ respectively, then the most probable value of the correlation between tests $i$ and $j$ would be $\sqrt{(p_i p_j)}$. He then showed that the correlations taken as a whole would give tetrad-differences of zero, in other words that the rank of the matrix excluding diagonal elements would be unity. 'There is', as he points out, 'nothing novel in the idea that correlations are due to a number of small causes. A similar idea is at the base of the usual treatment of the Gauss curve, and is found in Bravais. What is novel is the idea that correlations thus produced will tend to have zero tetrad-differences.'

Thomson further pointed out that the tendency towards *unit* rank in the matrix would exist only in as far as the samples of 'bonds' were drawn from the complete pool. Usually this will not be the case. Tests in a battery can often be classified at face value as *verbal*, *spatial*, *perceptual*, and so on. The tendency to unit rank would then be interfered with were it the case that these tests tended to sample from sub-pools of the bonds and this sampling would lead to 'group factors'. Thomson adds, "the correlations will be related *as if* they were due to a small number of common factors (no longer one only)..., but again from my point of view these 'factors' are not unitary entities. They arise naturally and mathematically from a theory of the overlap of many and much smaller realities."

The statistical aspects of Thomson's model were later elaborated by Bartlett (1937, 1953), who showed that from it emerged statistical entities with all the properties of general, group and specific factors. For a detailed account of the model and illustrations of its use in practice see Maxwell (1972a, b).

CHAPTER TWO

# General Observations

## Introduction

In this chapter some general problems are considered which are important in an appraisal of the multivariate models described in later chapters.

Because of wide criticism of Spearman's two-factor hypothesis, many of the salutory remarks which he made in his 1904 papers about the conduct of research in the psychological field tended to be overlooked. In particular he was deeply concerned about the reliability of the data that could be collected. Even though techniques of measurement have greatly improved since his day, a limiting feature of research work in the social sciences in general still is the fallibility of the observations on which it is based, a limitation which persists even in investigations which have been carefully designed and are based on adequate samples. The fallibility arises in part from the difficulty of defining psychological and social variables accurately, but more especially from the difficulty of obtaining precise measurements of them, once the effort to define them has been made. As a consequence, data presented for multivariate analysis are likely to be contaminated by errors of measurement. When employing a multivariate model it is well to be aware of this, especially if the model does not specifically make allowance for error. Some relevant points will now be noted.

## Estimating variances and covariances

Consider a random variate, say $Z$, which is assumed to be normally distributed in a given population. In the present context we might think of it as a test designed to measure the intelligence of eleven-year-old children in a country such as Britain. The distribution is defined by two population parameters, its mean $\mu$ and its variance $\sigma^2$. Although they are seldom known exactly it is helpful, in reconciling differing results for independent samples, to assume that true values of them exist. Given a random sample of $n$ observations, $z_i$, $(i = 1, \ldots, n)$ on the variate, estimates $\bar{z}$ and $s^2$ of $\mu$ and $\sigma^2$ respectively can be found by the expressions

$$\bar{z} = \Sigma z_i / n, \tag{2.1}$$

and

$$s^2 = \Sigma (z_i - \bar{z})^2 / (n - 1). \tag{2.2}$$

By statistical theory it can then be shown that

$$E(\bar{z}) = \mu \text{ and } E(s^2) = \sigma^2, \tag{2.3}$$

where $E$ stands for 'expected value of'. Hence the estimates $\bar{z}$ and $s^2$ are unbiased estimates of $\mu$ and $\sigma^2$ respectively. But there is a difficulty here which is seldom made explicit for underlying the statements in (2.3) is the assumption that the observations $z_i$ can be made without error, and this is seldom the case in practice. Let us thus make the more realistic assumption that the $z_i$ can only be determined imprecisely. The actual observations are $x_i$, given by

$$x_i = z_i + e_i, \qquad (i = 1, \ldots, n) \tag{2.4}$$

where the $e_i$ are errors. We shall suppose that the latter are independent of one another and of the $z_i$ and are normally distributed about zero with variance $\sigma_e^2$ in the population. Using the $n$ observations $x_i$, estimates $\bar{x}$ and $s_x^2$ of $\mu$ and $\sigma^2$ can be obtained. Since $E(e_i) = 0$ it follows, using (2.4), that

$$E(x_i) = E(z_i) = \mu,$$

so that $\bar{x}$ yields an unbiased estimate of the population mean. But $s_x^2$ is not an unbiased estimate of $\sigma^2$ for, in view of the variance of the errors in addition to that of the $z_i$,

$$E(s_x^2) = \sigma^2 + \sigma_e^2. \tag{2.5}$$

In other words an estimate of the true variance of a variate, based on observations which are subject to random error, is on average an over-estimate. Now in situations in which errors of measurement are known to be trivial in magnitude, so that $\sigma_e^2$ is small relative to $\sigma^2$, it would be pedantic to worry overmuch about the bias in an estimate of variance. But in research work in the social sciences, generally this is not the case. For example, in clinical investigations in psychology the several batteries of cognitive tests in the Wechsler series are widely used. For the subtests in these batteries it can be shown that as much as 20% of the observed variance may be error variance, and this is far from trivial. Since these tests are typical of the kinds of measuring instruments available to psychologists it follows that error variance presents a real problem and should be borne in mind when choosing a model. As an illustration, suppose that we wanted to examine the correlational structure of the subtests in one of the Wechsler batteries. It would now be expedient to choose the factor analysis model (see Chapter 4) rather than the principal component model (Chapter 3), since the former makes explicit allowance for error variance while the latter does not.

Our choice of the factor model is further supported by another elementary fact from estimation theory. It is that an estimate of the covariance of two random normal variates, obtained in the usual way from fallible observations, is unbiased and the factor model, which is 'covariance-orientated', uses this advantage to the full. To demonstrate the former point let us consider two variates $Z_1$ and $Z_2$ which have a bivariate normal distribution in the population. Let $x_1$ and $x_2$ represent a pair of scores on them for a member of the population and assume, for algebraic convenience, that the means of the variates are zero. In terms of 'true' and 'error' scores we may write

$$x_1 = z_1 + e_1 \text{ and } x_2 = z_2 + e_2. \tag{2.6}$$

The estimate of the covariance of $Z_1$ and $Z_2$ is then given by

$$\text{cov}(Z_1 Z_2) = \Sigma(x_1 x_2)/(n-1), \tag{2.7}$$

where summation is over the $n$ pairs of observations in a sample. The expected value of the covariance of the variates is tentatively written as $E(x_1\, x_2)$. Using the expressions in (2.6), we get

$$E(x_1 x_2) = E\{(z_1 + e_1)(z_2 + e_2)\},$$

which reduces to

$$E(x_1 x_2) = E(z_1 z_2) \tag{2.8}$$

on the assumption that errors and true scores are independent of each other. It follows that (2.7) provides an unbiased estimate of the covariance of the variates despite the fact that the $x$'s are subject to errors of measurement. As we shall see in Chapter 4, this convenient fact may enable us to obtain unbiased estimates of variances of variates even in cases where the observations cannot be made with precision, and so to estimate the amount of error variance involved.

**Linear constraints on variates**

Let $Z_j$ represent a set of $p$ random variates ($j = 1, \ldots, p$). The variates are said to be linearly independent if no member of the set in an exact multiple of another, or can be expressed as a linear sum of some or all of the others. However, linear independence does not rule out the possibility that the variates may be correlated. Correlation between two linearly independent variates will arise if either consists in part of the other, in the sense implied is a simple regression equation, the remaining part or residual being uncorrelated with the other.

Assume that the true variances and covariances of the variates are known and are arranged in a square matrix of order $p$, the variances occupying the cells in its main diagonal. When the variates are linearly independent the variance-covariance matrix (often referred to simply as the covariance or dispersion matrix) is said to be of full rank (see Chapter 3), that is of rank $p$. Moreover, the matrix retains its full rank if it is transformed into a correlation matrix. Its main diagonal elements are then unities while the correlation between any pair of the variates is obtained by dividing their covariance by the square root of the product of their variances.

Now, under certain conditions it is possible to reduce the rank of a matrix, without altering its order $p$, by reducing the values of its diagonal elements, a procedure which is reasonable when these contain a considerable margin of variance due to error. As an example, consider the matrix

$$\begin{bmatrix} 1.00 & 0.42 & 0.35 & 0.28 \\ 0.42 & 1.00 & 0.30 & 0.24 \\ 0.35 & 0.30 & 1.00 & 0.20 \\ 0.28 & 0.24 & 0.20 & 1.00 \end{bmatrix},$$

which we may think of as a correlation matrix for four variates. Examination of it will show that it is of full rank, since none of its columns or rows can be derived as a linear sum of the others. None-the-less the matrix, if we omit its diagonal entries of unity, has a particular structure, for on choosing any minor determinant of order two from it, for example the minor –

$$\begin{vmatrix} 0.42 & 0.28 \\ 0.30 & 0.20 \end{vmatrix},$$

and on applying equation (1.4) its value is found to be zero. In detail,

$$0.42 \times 0.20 - 0.30 \times 0.28 = 0.$$

If the diagonal elements of the matrix are now adjusted so that they too fit in with the criterion of zero minors, the matrix becomes

$$\begin{bmatrix} 0.49 & 0.42 & 0.35 & 0.28 \\ 0.42 & 0.36 & 0.30 & 0.24 \\ 0.35 & 0.30 & 0.25 & 0.20 \\ 0.28 & 0.24 & 0.20 & 0.16 \end{bmatrix}.$$

This matrix, though of order four has a rank only of unity, for the entries in any of its rows (or columns) is a multiple of the entries in any other row (or column). For example, the second row can be derived from the first by multiplying the elements in the latter by the constant 6/7.

In this instance the reduction in the rank of the matrix is possible because of constraits on its non-diagonal elements. Indeed, as we saw in Chapter 1, these are the constraits which would need to be fulfilled for Spearman's two-factor hypothesis to be true, though with real data one could only expect the values of the minor determinants to be zero within the limits of sampling error.

## Matrices of reduced rank

Except in an item analysis in which the variates are items of homogenous content and average difficulty, (say items from a cognitive test, or other similar measuring device) matrices in which the correlations are approximately of unit rank, are rarely found in practice. But it is now well established, and may be taken as an empirical fact, that correlation matrices derived from psychological and other such-like variates, invariably can be shown to have a rank less than their order. The details of how this is done, and of the statistical tests required to varify the claim, are left to Chapter 5. Here we will mention a few of the reason which could account for the finding, though in doing so it is difficult to avoid an accusation of circular argument. An elementary point is that a battery of tests assembled for a given investigation may contain subsets of tests bearing a strong family relationship to each other, such as *verbal* tests or *numerical* tests. Within each subset, as in the case of an item analysis, a tendency to unit rank in their intercorrelations is likely to exist and this will contribute towards a reduction of the rank of the overall matrix. But the basic reasons for the reduction in rank are probably physiological. Three points of common knowledge may be noted. The first concerns the tendency towards localization of function in the brain, which of itself must lead to some classification of the tasks which individuals perform. The second is that the number of nerve cells in the brain is insufficient to suppose that a unique set of them is available for every simple memory stored and every simple conditioned reflex which is learned. This may be a basic 'cause' of the correlations observed between tests of mental performance and, when taken in conjunction with the first point, of the differential correlation between different classes of test. The third point is concerned with redundancy in the brain, for it is very unlikely that exactly the same neurons are involved in each repetition of even a simple mental task. We recall that Thomson in his model pressed the latter point to its extreme limit, when he postulated that the performance of a given test involved a *random sample* of particular size of basic elements of the brain. Under this assumption he was able to show that the intercorrelations of a set of tests would be constrained to have unit rank. Bartlett extended the notion to the case where particular subsets of like-tests were assumed to sample selectively from the basic elements, and indicated that the rank of the correlations as a whole would then be increased by unity for each subset. To the extent then that redundancy in the brain, as here interpreted, does exist and to the extent that the brain's basic units are shared in the performance of mental tasks and localization of function exists, one would expect the intercorrelations for a set of mental tests to have a rank lower than the number of tests in the set. This is the most one can say in the light of present knowledge, but the contention is strongly supported by observation.

## Terminology

Following Galton's postulation of a continuum along which people could be placed with respect to their intelligence, it became common practice in psychological work to postulate hypothetical variates, or dimensions as they are commonly called, which when objectively defined could be employed in the description of human behaviour. Now the word 'dimension' has a rigid geometric connotation and, thought of in this way, seems an inappropriate term to use in a psychological context. However, it can be given an 'as if' interpretation which makes it more appropriate and acceptable. This is seen for instance in the case of Thomson's model of the brain, in which a set of correlation coefficients having unit rank arose as a consequence of sampling from a very large number of small elements. But, once obtained, the unit rank might, for convenience of discussion, be described in dimensional terms, 'as if' it were due to a single hypothetical variate shared by the original variates which (as Spearman postulated) was the cause of the correlations between them.

## Metric

Many of the classic models in multivariate analysis, such as the multiple regression, principal component, canonical correlation and factor analysis models, are applied to correlation rather than to covariance matrices. This is so primarily because most of the variates employed by social scientists have arbitrary scales, and since their true relative variances are unknown it is expedient to equate them. This is done by expressing the observed scores on each variate in standardised form. For example, if $x_i (i = 1, \ldots, n)$ is a sample of $n$ scores on a variate $X$ their standardized equivalents are given by

$$y_i = (x_i - \bar{x})/s,$$

where $\bar{x}$ is the sample mean and $s$ the sample standard deviation. If, for a number of variates, standardized scores rather than raw scores are employed then their covariance matrix, calculated in the usual way, is their correlation matrix.

But the procedure has limitations which need to be borne in mind. For example, if samples from two or more populations are being compared and if the several variates being measured have different variances in each population then the use of their correlation rather than their covariance matrices is likely to lead to confusion. This will occur in the case of a multivariate technique which is not invariant under changes of scale in the variates. For example, the weights obtained in a principal component analysis when a correlation matrix is used bear no direct relationship to those obtained when the corresponding covariance matrix is employed. As a consequence, information about differences in variances in different populations once discarded cannot be retrieved.

In other situations, different considerations may determine our choice of scaling. For example, in canonical correlation analysis it is customary to use the correlation rather than the covariance matrix, otherwise the weights derived may be very difficult to interpret. In the case of canonical variate analysis, freedom of choice exists as the *scores* on the canonical variates are invariant under change of scale in the observed variates. The calculations may most conveniently and most accurately be performed if the variates are now scaled so as to have variances of the same order of magnitude.

Special problems of metric arise too in the analysis of dichotomously scored variates, the only admissible scores being zero and unity. For such variates the means are proportions, where $p_j$ is the proportion of individuals in a sample with scores of unity on the $j$th variate. The variance of the variate is $p_j q_j$, where $q_j = 1 - p_j$, and it is easy to show that variates with $p$-values lying in the range 0.2 to 0.8 have variances which do not differ very much in magnitude. In view of this it is customary to employ a correlation matrix when investigating the interrelationships of such variates. Almost invariably phi-coefficients rather than tetrachoric correlation coefficients are calculated for they can be found by the ordinary product-moment formula and computer programs are readily available. However, phi-coefficients have the disadvantage that estimates of them are not independent of the means of the variates when the latter are unequal. As a consequence, a spurious factor emerges when a matrix of phi-coefficients is subjected to a principal component or factor analysis. If the matrix is of very low rank the spurious factor can generally be detected immediately as it will have weights roughly proportional to the $p$-values of the variates. In other cases it may be necessary to rotate the factors to reveal its full contribution. The alternative is to eliminate it from the correlation matrix before a component analysis is embarked upon (Lawley, 1944). See also the final section of Chapter 4.

### The distribution of variates

For most of the models to be considered it is assumed that the variates involved have a multivariate normal distribution in the population sampled. This implies that each of the variates is normally distributed and, amongst other things, that the regression of any variate on any other, or on any set of the others, is linear.

Whilst the overall assumption of multivariate normality is very difficult to test, partial evidence of its truth can readily be obtained by examining the variates individually for normality and in pairs for bivariate normality. At this stage too transformations of variates which would render them 'normal', if they are found not to be so, should be considered.

The main advantage of multivariate normality is that it allows relatively straightforward models to be formulated for the analysis of data. A secondary advantage is that it ensures in general that the population sampled is homogeneous with respect to the variates in question, that is that it is not composed of relatively distinct sub-populations which, if undetected, would give rise to bias in estimating means, variances and other summary statistics and could lead to invalid conclusions being drawn from a model.

CHAPTER THREE

# Matrices and Determinants

## Matrices and their manipulation

In the preceding chapter the word 'matrix' has been frequently used. It refers to a rectangular or square array of numbers or symbols. Such arrays occur constantly in multivariate analysis and it is convenient to have a concise symbolism for dealing with them. This is provided by the theory of matrix algebra. Some elementary ideas from that theory are provided in this chapter and these are used extensively in succeeding chapters.

It is customary to use capital boldface letters to denote matrices whose elements are real numbers. For example we may write:

$$\mathbf{A} = \begin{bmatrix} a_{11} & a_{12} & \cdot & \cdot & \cdot & a_{1p} \\ a_{21} & a_{22} & \cdot & \cdot & \cdot & a_{2p} \\ \cdot & \cdot & \cdot & \cdot & \cdot & \cdot \\ \cdot & \cdot & \cdot & \cdot & \cdot & \cdot \\ \cdot & \cdot & \cdot & \cdot & \cdot & \cdot \\ a_{p1} & a_{p2} & \cdot & \cdot & \cdot & a_{pp} \end{bmatrix}$$

or more concisely $\mathbf{A} = [a_{ij}]$, where $i$ and $j$ run from 1 to $p$. In this symbolic form $i$ refers to rows of the matrix and $j$ to columns. The above matrix is square and its north-west to south-east disgonal, in which it is seen that the subscripts $i$ and $j$ are equal, is called the main diagonal of the matrix. If, in a square matrix, $a_{ij} = a_{ji}$ the matrix is said to be symmetric about its main diagonal: this is the case for a matrix of correlations.

**Addition.** Two matrices **A** and **B**, which have the same number of rows and columns can be added by adding corresponding elements. For example

$$\text{if } \mathbf{A} = \begin{bmatrix} 4 & 2 \\ 3 & 5 \end{bmatrix} \text{ and } \mathbf{B} = \begin{bmatrix} 1 & 3 \\ 4 & 7 \end{bmatrix}, \text{ then } \mathbf{A} + \mathbf{B} = \begin{bmatrix} 5 & 5 \\ 7 & 12 \end{bmatrix}.$$

This result can be written symbolically as:

$$\mathbf{A}+\mathbf{B}=[a_{ij}]+[b_{ij}]=[a_{ij}+b_{ij}]$$

The product of a matrix and a real number, or scalar, $k$ is defined by

$$k\mathbf{A}=\mathbf{A}k=[ka_{ij}],$$

for example, if

$$\mathbf{A}=\begin{bmatrix}4 & 2\\ 3 & 5\end{bmatrix} \text{ and } k=2, k\mathbf{A}=2\begin{bmatrix}4 & 2\\ 3 & 5\end{bmatrix}=\begin{bmatrix}8 & 4\\ 6 & 10\end{bmatrix}.$$

A *null matrix* is one whose elements are all zero and we may then write $\mathbf{A}=0$. A *diagonal matrix* is one with zeros in all cells except the main diagonal.
For example

$$\mathbf{D}=\begin{bmatrix}3 & & \\ & 2 & \\ & & 4\end{bmatrix}$$

is a diagonal matrix.
If in a diagonal matrix all the diagonal entries are unity the matrix is called the *identity matrix* and is denoted by $\mathbf{I}$; this corresponds to *unity* in ordinary algebra.

**Rectangular matrices.** Matrices in which the number of rows is not equal to the number of columns are called rectangular matrics. Here is an example

$$\mathbf{B}=\begin{bmatrix}4 & 7\\ 3 & 1\\ 5 & 2\end{bmatrix}.$$

If a matrix is re-written with its rows as columns the matrix is said to be transposed. This is frequently indicated by the use of a *prime* so that the transpose of $\mathbf{B}$ is denoted by $\mathbf{B}'$. In other words

$$\mathbf{B}'=\begin{bmatrix}4 & 3 & 5\\ 7 & 1 & 2\end{bmatrix}.$$

If a matrix is square and symmetric it follows that its transpose is identical to the matrix itself.

**Row and column vectors.** A matrix consisting of a single row is referred to as a *row vector*. Vectors are denoted by lower case boldface letters. An example of a column vector is:

$$\mathbf{c} = \begin{bmatrix} 3 \\ 5 \\ 1 \end{bmatrix}.$$

The transpose of **c**, namely **c**′, is a *row vector*, namely

$$\mathbf{c}' = [3 \quad 5 \quad 1].$$

To save space column vectors are frequently denoted by curly brackets so that we may write $\mathbf{c} = \{3 \ \ 5 \ \ 1\}$.

**Matrix multiplication.** If we take the row vector $[a \quad b \quad c]$ and the column vector $\{x \quad y \quad z\}$, multiply corresponding terms together and add the results we obtain $(ax + by + cz)$. This is defined as the *product* of the two vectors. Written in full we have:

$$[a \quad b \quad c] \begin{bmatrix} x \\ y \\ z \end{bmatrix} = (ax + by + cz).$$

An extension of this rule enables us to find the product of two matrices say **A** and **B**. Let

$$\mathbf{A} = \begin{bmatrix} 4 & 2 & 3 \\ 5 & 1 & 2 \end{bmatrix} \text{ and } \mathbf{B} = \begin{Bmatrix} 2 & 1 & 4 \\ 3 & 5 & 2 \end{Bmatrix},$$

then

$$\mathbf{AB} = \begin{bmatrix} 4 & 2 & 3 \\ 5 & 1 & 2 \end{bmatrix} \begin{bmatrix} 2 & 3 \\ 1 & 5 \\ 4 & 2 \end{bmatrix} = \begin{bmatrix} 22 & 28 \\ 19 & 24 \end{bmatrix}.$$

In obtaining the answer we follow the rule for vector multiplication given above. For example

22 is the product of $[4 \quad 2 \quad 3]$ and $\{2 \quad 1 \quad 4\}$,

28 is the product of $[4 \quad 2 \quad 3]$ and $\{3 \quad 5 \quad 2\}$,

19 is the product of $[5 \quad 1 \quad 2]$ and $\{2 \quad 1 \quad 4\}$,

and 24 is the product of $[5 \quad 1 \quad 2]$ and $\{3 \quad 5 \quad 2\}$.

In other words the first row of **A** is multiplied in on each column of **B** to give the first row of the answer: the second row of **A** is multiplied in on each column of **B** to give the second row of the answer. Notice that for the multiplication to be possible the matrix placed first must have as many columns as that placed second has rows. The product **BA** can also in this instance be found and then we obtain:

$$\mathbf{BA} = \begin{bmatrix} 23 & 7 & 12 \\ 29 & 7 & 13 \\ 26 & 10 & 16 \end{bmatrix}.$$

Hence $\mathbf{AB} \neq \mathbf{BA}$. In this respect matrix algebra differs from ordinary algebra: in matrix algebra we must distinguish between pre-multiplication and post-multiplication.

The student may verify for himself using simple arithmetic examples, the following rules of matrix algebra. If **A**, **B**, **C**, are matrices of order $p \times q$, $q \times r$, $r \times s$, then:

$$\mathbf{A}(\mathbf{BC}) = (\mathbf{AB})\mathbf{C} = \mathbf{ABC}.$$

Matrix multiplication also satisfies the laws

$$\mathbf{A}(\mathbf{B} + \mathbf{C}) = \mathbf{AB} + \mathbf{AC},$$

and

$$(\mathbf{A} + \mathbf{B})\mathbf{C} = \mathbf{AC} + \mathbf{BC}.$$

If **A** is a square matrix

$$\mathbf{A}^2 = \mathbf{AA}.$$

Moreover, using the fact that the transpose of $\mathbf{A} = [a_{ij}]$ is $\mathbf{A}' = [a_{ji}]$, with similar expression for **B** and **C**, it can be shown that:

$$(\mathbf{AB})' = \mathbf{B}'\mathbf{A}',$$

$$(\mathbf{ABC})' = \mathbf{C}'\mathbf{B}'\mathbf{A}',$$

and so on.

**Determinants**

Associated with any square matrix, whether or not it is symmetric, is a determinant. The determinant of the matrix $\mathbf{A} = [a_{ij}]$ is written $|\mathbf{A}| = |a_{ij}|$. Unlike matrices, which are simply arrays of numbers, determinants can be evaluated and have actual numerical values if the terms $a_{ij}$ are numerical, or stand for algebraic expressions if these terms are algebraic symbols. The simplest example is a $2 \times 2$ determinant and the expression $\begin{vmatrix} a & b \\ c & d \end{vmatrix}$ is an alternative way of writing $(ad - bc)$. Similarly

$$\begin{vmatrix} 2 & 4 \\ 1 & 7 \end{vmatrix} = (2 \times 7) - (4 \times 1) = 10.$$

All larger determinants are evaluated in terms of determinants of order $2 \times 2$ (also known as *minor* determinants of order 2). For example the $3 \times 3$ determinant:

$$|\mathbf{A}| = \begin{vmatrix} a & b & c \\ d & e & f \\ g & h & i \end{vmatrix}, \text{ can be written as}$$

$$= a\begin{vmatrix} e & f \\ h & i \end{vmatrix} - b\begin{vmatrix} d & f \\ g & i \end{vmatrix} + c\begin{vmatrix} d & e \\ g & h \end{vmatrix},$$

$$= aei - ahf - bdi + bgf + cdh - cge.$$

The rule for expanding a determinant can easily be deduced by inspection of this example. One row (or column) is chosen: in our example the first row. The terms in this row are multiplied in order by $(-1)^{i+j} = 1, -1, 1, \ldots$ and these are used as multipliers for determinants of order one less than the original. In our example when the term $a$ is chosen the minor determinant following it is obtained by deleting from $|\mathbf{A}|$ the row and column in which $a$ stands. A similar procedure is followed for each term in the row (or column) chosen.

A determinant when evaluated may have the value zero. For example,

$$|\mathbf{A}| = \begin{vmatrix} 2 & 4 \\ 3 & 6 \end{vmatrix} = 12 - 12 = 0.$$

As we shall shortly see, the rank of a matrix can be defined in terms of the determinants of given order which are zero.

## Inverse matrices

So far we have discussed the addition and multiplication of matrices and we now come to the matrix equivalent of division. In ordinary algebra the inverse of a number $a$ is written as $a^{-1}$, and we know that $a \times a^{-1} = 1$. In a corresponding manner the inverse of a matrix $\mathbf{A}$ is written $\mathbf{A}^{-1}$ and when the two are multiplied we write

$$\mathbf{AA}^{-1} = \mathbf{I},$$

where $\mathbf{I}$ is the identity matrix, that is a matrix with unities in its main diagonal and zeros in all the other cells.

Only square matrices have inverses. Obtaining the inverse of a matrix of order greater than 5 or 6 involves laborious calculations. Now-a-days these can be performed very speedily on electronic computers so that all we require here is an understanding of the properties of inverses. For the simplest case in which

$$\mathbf{A} = \begin{bmatrix} a & b \\ c & d \end{bmatrix}, \mathbf{A}^{-1} = \frac{1}{|A|}\begin{bmatrix} d & -b \\ -c & a \end{bmatrix} \tag{3.1}$$

In numerical terms if

$$\mathbf{A} = \begin{bmatrix} 4 & 3 \\ 1 & 2 \end{bmatrix}, \text{ then } \mathbf{A}^{-1} = \frac{1}{5}\begin{bmatrix} 2 & -3 \\ -1 & 4 \end{bmatrix}.$$

The result can be verified by multiplication. Hence

$$\mathbf{AA}^{-1} = \begin{bmatrix} 4 & 3 \\ 1 & 2 \end{bmatrix}\begin{bmatrix} 2/5 & -3/5 \\ -1/5 & 4/5 \end{bmatrix} = \begin{bmatrix} 1 & 0 \\ 0 & 1 \end{bmatrix} = \mathbf{I}.$$

The reader may verify that $\mathbf{AA}^{-1} = \mathbf{A}^{-1}\mathbf{A}$, and that if $\mathbf{A}$, $\mathbf{B}$ and $\mathbf{C}$ are all square non-singular (defined below) matrices of the same order

$$(\mathbf{ABC})^{-1} = \mathbf{C}^{-1}\mathbf{B}^{-1}\mathbf{A}^{-1} \tag{3.2}$$

Since the inverse of the matrix $\mathbf{A}$ involves the multiplier $1/|\mathbf{A}|$ it is immediately clear that the process will break down if $|\mathbf{A}| = 0$, for this would lead to dividing by zero, which in algebra is not allowed. When the determinant of a matrix is zero the matrix is said to be *singular* and singular matrices do not have inverses in the ordinary sense. This fact, together with some other properties of matrices and determinants, can be demonstrated in a vivid way by applying the theory so far presented to the problem of solving simple linear equations.

## Some uses of matrices and determinants

Take the two simultaneous equations

$$2x + 3y = 7$$

$$3x - y = 5.$$

If the second equation is multiplied by 3 and the result added to the first equation we get $11x = 22$, or $x = 2$. Substituting this value in either equation gives $y = 1$, hence the solutions to the equations are $x = 2$, $y = 1$.

Now the two equations may be written in matrix form as:

$$\begin{bmatrix} 2 & 3 \\ 3 & -1 \end{bmatrix}\begin{bmatrix} x \\ y \end{bmatrix} = \begin{bmatrix} 7 \\ 5 \end{bmatrix},$$

or, in symbolic notation, as

$$\mathbf{Az} = \mathbf{b}, \tag{3.3}$$

where

$$\mathbf{z} = \begin{bmatrix} x \\ y \end{bmatrix}, \quad \mathbf{A} = \begin{bmatrix} 2 & 3 \\ 3 & -1 \end{bmatrix} \quad \text{and } \mathbf{b} = \begin{bmatrix} 7 \\ 5 \end{bmatrix}.$$

Premultiplying (3.3) by $\mathbf{A}^{-1}$ we get

$$\mathbf{z} = \mathbf{A}^{-1}\mathbf{b} \tag{3.4}$$

Using (3.1) we find

$$\mathbf{A}^{-1} = -\frac{1}{11}\begin{bmatrix} -1 & -3 \\ -3 & 2 \end{bmatrix} = \frac{1}{11}\begin{bmatrix} 1 & 3 \\ 3 & -2 \end{bmatrix}.$$

Substituting the value of $\mathbf{A}^{-1}$ and of $\mathbf{b}$ in (3.4) gives

$$\mathbf{z} = \frac{1}{11}\begin{bmatrix} 1 & 3 \\ 3 & -2 \end{bmatrix}\begin{bmatrix} 7 \\ 5 \end{bmatrix} = \frac{1}{11}\begin{bmatrix} 22 \\ 11 \end{bmatrix} = \begin{bmatrix} 2 \\ 1 \end{bmatrix}.$$

And since $\mathbf{z} = \begin{bmatrix} x \\ y \end{bmatrix}$, we have $x = 2$ and $y = 1$, as required.

We now have seen that simultaneous linear equations can be solved by matrix algebra. The advantages of this are far from obvious in the simple example just considered, but happily the matrix method can be applied in a routine way to the solution of linear equations involving many variables and the student who remembers the laborious processes involved in solving such equations by ordinary algebraic methods will appreciate the point.

Now let us try to solve the equations

$$3x + 2y = 4,$$
$$15x + 10y = 20.$$

If the first equation is multiplied by 5 and the second equation is subtracted from the result all terms disappear; the equations cannot be solved. The matrix procedure given above makes this point clear immediately, for if we let

$$\mathbf{A} = \begin{bmatrix} 3 & 2 \\ 15 & 10 \end{bmatrix}$$

we see that $|\mathbf{A}| = 0$, hence $\mathbf{A}$ is singular and $\mathbf{A}^{-1}$ cannot be found.

In the last example the equations could not be solved because they are not independently true: the second is simply a multiple of the first and so does not provide the additional information required for a solution. To test if the equations in a set are independent it is only necessary to set up the determinant of the coefficients of the variables, $x$, $y$, etc., and show that it is not equal to zero. For the above equations the determinant, as we have seen, is

$$\begin{vmatrix} 3 & 2 \\ 15 & 10 \end{vmatrix},$$

which is zero.

## Rank of matrix

Let us now look at the following three equations:

$$\begin{aligned} 2x + y - 5 &= 0, \\ x - 2y &= 0, \\ 3x - y - 5 &= 0, \end{aligned} \tag{3.5}$$

all three are satisfied by the values $x = 2$ and $y = 1$, in other words the three equations are consistent. This fact can be demonstrated by showing that the determinant $|\mathbf{A}|$ of the matrix:

$$\mathbf{A} = \begin{bmatrix} 2 & 1 & -5 \\ 1 & -2 & 0 \\ 3 & -1 & -5 \end{bmatrix} \tag{3.6}$$

is zero.

Now since only two of the original three equations are necessary for solving for $x$ and $y$ the matrix $\mathbf{A}$ contains redundant information: its rows (or columns) are not all independent.

The number of independent rows (or columns) of a matrix is called the *rank* of the matrix. One way of determining the rank of a matrix is by evaluating determinants derived from it. A matrix $\mathbf{A}$

of order $p$ has rank $p$ if $|\mathbf{A}| \neq 0$. If $|\mathbf{A}|$ is zero then the matrix is of rank $(p-1)$ provided at least one of the minor determinants of order $(p-1)$ derived from it is non-zero. If all its minor determinants of order $(p-1)$ are zero then the matrix is of rank $(p-2)$ provided at least one of its minors of order $(p-2)$ is non-zero, and so on.

Take the matrix $\mathbf{A}$ given in (3.6) which is of order $p=3$. Expanding its determinant in terms of the *second* row we get

$$-1\begin{vmatrix} 1 & -5 \\ -1 & -5 \end{vmatrix} - 2\begin{vmatrix} 2 & -5 \\ 3 & -5 \end{vmatrix} = 10 - 2(5) = 0.$$

Hence $|\mathbf{A}| = 0$ and the matrix is of rank less than three. But in the expansion it is seen that its minors, of order $(p-1)=2$, are not all zero hence the matrix is of rank 2.

## Latent roots and vectors of a matrix

Consider the following square matrix, of order three, namely

$$\mathbf{A} = \begin{bmatrix} 11 & -6 & 2 \\ -6 & 10 & -4 \\ 2 & -4 & 6 \end{bmatrix}. \tag{3.7}$$

By expansion its determinant, $|\mathbf{A}|$, is found to be 324, which is non-zero, hence the matrix is of rank three. Now if a scalar quantity, $\lambda$, can be found such that the determinant

$$\begin{vmatrix} (11-\lambda) & -6 & 2 \\ -6 & (10-\lambda) & -4 \\ 2 & -4 & (6-\lambda) \end{vmatrix} \tag{3.8}$$

is zero, the rank of the resulting matrix will be only of rank two. The scalar quantity $\lambda$ is then said to be a *latent root* (also called *characteristic root* or *eigenvalue*) of the matrix $\mathbf{A}$. One value of $\lambda$ which meets this requirement is 18: but since the matrix is of rank three there are two others; they are in fact 6 and 3. These three values of $\lambda$ are the three latent roots of the matrix and it may be noted that their sum, which is 27, is equal to the sum of the terms in the main diagonal of $\mathbf{A}$ – what is called the *trace* of the matrix. Moreover, the value of $|\mathbf{A}|$ is given by the product of the latent roots, for example $324 = 18 \times 6 \times 3$.

The determinant in expression (3.8) can be written as the difference of two determinants, namely,

$$\begin{vmatrix} 11 & -6 & 2 \\ -6 & 10 & -4 \\ 2 & -4 & 6 \end{vmatrix} - \begin{vmatrix} \lambda & & \\ & \lambda & \\ & & \lambda \end{vmatrix},$$

or, symbolically, as

$$|\mathbf{A} - \lambda\mathbf{I}|.$$

Hence the problem of finding the latent roots of a matrix **A** can be expressed as that of solving the determinantal equation

$$|\mathbf{A} - \lambda\mathbf{I}| = 0. \tag{3.9}$$

**Latent vectors**

If **A** is a square matrix, **x** a column vector and $\lambda$ is a scalar quantity such that

$$\mathbf{A}\mathbf{x} = \lambda\mathbf{x}, \tag{3.10}$$

then **x** is said to be a *latent vector* of the matrix **A**. Equation (3.10) can also be written as

$$(\mathbf{A} - \lambda\mathbf{I})\mathbf{x} = 0. \tag{3.11}$$

Alternatively, if **x** is a row vector the equation takes the form

$$\mathbf{x}'(\mathbf{A} - \lambda\mathbf{I}) = 0. \tag{3.12}$$

Now the condition necessary for equation (3.11) or (3.12) to be true is that the determinant $|\mathbf{A} - \lambda\mathbf{I}|$ be zero. But, as is clear from equation (3.9), this is the condition required for $\lambda$ to be a latent root of the matrix **A**. It can be shown that for a given value of $\lambda$, satisfying $|\mathbf{A} - \lambda\mathbf{I}| = 0$, there exists a vector $x$ satisfying the equation (3.10). If the matrix **A** is symmetric then, for a given $\lambda$, the column vector in equation (3.11) is the transpose of the row vector in equation (3.12). If **A** is asymmetric this will not be the case.

Finding the latent roots and vectors of a matrix involves laborious calculations but computer programs are readily available. Only in the case of a $2 \times 2$ matrix can they readily be found by hand. As an example, suppose that we require the latent roots and vectors of the matrix

$$\begin{bmatrix} 6 & 3 \\ 3 & 4 \end{bmatrix}.$$

Using (3.8) we set up the determinant

$$\begin{vmatrix} 6-\lambda & 3 \\ 3 & 4-\lambda \end{vmatrix}.$$

On expanding it and setting the result equal to zero, we get

$$(6-\lambda)(4-\lambda)-9=0,$$

which gives the quadratic equation

$$\lambda^2-10\lambda+15=0.$$

The roots of this equation are $\lambda_1=8.162$ and $\lambda_2=1.838$. To find the latent vector, say $\{x \quad y\}$, corresponding to $\lambda_1$, we use (3.10) and set up the matrix equation

$$\begin{bmatrix} 6 & 3 \\ 3 & 4 \end{bmatrix}\begin{bmatrix} x \\ y \end{bmatrix}=8.162\begin{bmatrix} x \\ y \end{bmatrix},$$

which yields the two linear equations

$$6x+3y=8.162x,$$

$$3x+4y=8.162y.$$

To solve these we set either $x$ or $y$ equal to unity and choose the solution which gives an absolute value of the other less than unity. Setting $x=1$ we obtain, from either equation,

$$y=0.721,$$

which is less than unity in absolute value, as required. Hence the latent column vector corresponding to the latent root 8.162 has elements $x=1$ and $y=0.721$. In a similar way the elements of the latent vector corresponding to $\lambda_2$ may be found and we get $x=-0.721$ and $y=1$.

For the matrix **A** given above the latent column vectors corresponding to its three latent roots 18, 6 and 3, which satisfy (3.10), are shown in the respective columns of the matrix:

$$\mathbf{X}=\begin{bmatrix} 1.0 & 1.0 & 0.5 \\ -1.0 & 0.5 & 1.0 \\ 0.5 & -1.0 & 1.0 \end{bmatrix}.$$

Generally a computer program gives each in normalized form. A vector is said to be normalised when the sum of the squares of its elements equals unity. To normalize the first column vector in **X** we find the sum of the squares of its elements, which is 2.25, and then divide each element by $\sqrt{2.25}$. When all the columns are treated in this way we get the matrix:

$$\mathbf{Z} = \begin{bmatrix} 0.667 & 0.667 & 0.333 \\ -0.667 & 0.333 & 0.667 \\ 0.333 & -0.667 & 0.667 \end{bmatrix}.$$

These numerical values may now be used to verify some general results in martix algebra. If $\mathbf{A}$ is a square symmetric matrix of order and rank $p$, $\mathbf{D}$ is a diagonal matrix of the same order whose elements are the latent roots of $\mathbf{A}$ and $\mathbf{Z}$ is a matrix whose columns are the normalized vectors corresponding to the successive latent roots, then

$$\mathbf{Z'Z} = \mathbf{I}, \tag{3.13}$$

where $\mathbf{Z'}$ is the transpose of $\mathbf{Z}$, and

$$\mathbf{A} = \mathbf{ZDZ'}. \tag{3.14}$$

Moreover, if the latter equation is pre-multiplied, on both sides, by $\mathbf{Z'}$ and post-multiplied by $\mathbf{Z}$ we obtain, using (3.13), the equation

$$\mathbf{Z'AZ} = \mathbf{D}. \tag{3.15}$$

In addition it can be verified that

$$|\mathbf{A}| = |\mathbf{Z}||\mathbf{D}||\mathbf{Z'}|, \tag{3.16}$$

and that

$$|\mathbf{Z}| = |\mathbf{Z'}| = 1, \tag{3.17}$$

from which it follows that

$$|\mathbf{A}| = |\mathbf{D}|, \tag{3.18}$$

Since $|\mathbf{D}|$ is equal to the product of the latent roots then so is $|\mathbf{A}|$.

## Orthogonal matrices

A matrix $\mathbf{B}$ is said to be orthogonal (orthonormal) if

$$\mathbf{BB'} = \mathbf{B'B} = \mathbf{I}. \tag{3.19}$$

From this it follows that for an orthogonal matrix:

$$\mathbf{B'} = \mathbf{B}^{-1}, \text{ and } \mathbf{B} = (\mathbf{B'})^{-1} \tag{3.20}$$

An example is matrix $\mathbf{Z}$ above. Using (3.2) and (3.20) it can easily be shown that if the latent roots and vectors of a matrix are known its inverse can readily be found. For example, from expression (3.14) we have $\mathbf{A} = \mathbf{ZDZ}'$ where $\mathbf{A}$ is symmetric. Hence,

$$\begin{aligned} \mathbf{A}^{-1} &= \mathbf{Z}'^{-1}\mathbf{D}^{-1}\mathbf{Z}^{-1}, \qquad \text{using (3.2)} \\ &= \mathbf{ZD}^{-1}\mathbf{Z}', \qquad \text{using (3.20)}, \end{aligned} \tag{3.21}$$

in which the elements of $\mathbf{D}^{-1}$ are $1/\lambda_i$. Note too that

$$\begin{aligned} \mathbf{A}^2 = \mathbf{AA} &= (\mathbf{ZDZ}')(\mathbf{ZDZ}') \\ &= \mathbf{ZD}^2\mathbf{Z}'. \end{aligned} \tag{3.22}$$

## Matrix rotation

Rotation of axes is a common procedure in principal component and factor analysis. If a point P has co-ordinates $(x_1, x_2)$ with reference to a given set of orthogonal axes and if the axes are rotated about the origin through an angle $\theta$ in a clockwise direction the co-ordinates $(y_1, y_2)$ of P with reference to the rotated axes are obtained by postmultiplying the row vector $x' = [x_1 \quad x_2]$ by an orthogonal matrix, say

$$\mathbf{U} = \begin{bmatrix} \cos\theta & \sin\theta \\ -\sin\theta & \cos\theta \end{bmatrix}. \tag{3.23}$$

Denoting the row vector $[y_1 \quad y_2]$ by $\mathbf{y}'$, we have

$$\mathbf{y}' = \mathbf{x}'\mathbf{U},$$

hence

$$\mathbf{y}'\mathbf{y} = \mathbf{x}'(\mathbf{UU}')\mathbf{x} = \mathbf{x}'\mathbf{x}, \tag{3.24}$$

since $\mathbf{UU}' = \mathbf{I}$. Now $\mathbf{x}'\mathbf{x} = x_1^2 + x_2^2$ is the square of the distance of P from the origin, and it follows from (3.24) that this distance is unaffected by the rotation.

## Triangular matrices

If, for a matrix $\mathbf{T} = [t_{ij}]$, $t_{ij} = 0$ for $i < j$, then all the elements above the diagonal are zero. $\mathbf{T}$ is then said to be a lower triangular matrix: its transpose $\mathbf{T}'$ is upper triangular. The product of two lower triangular matrices is also lower triangular, and of two upper triangular matrices is upper triangular. $|\mathbf{T}|$ is given simply by $(t_{11}t_{22} \ldots t_{pp})$, that is by the product of the diagonal elements of $\mathbf{T}$.

The inverse of a lower triangular matrix is also lower triangular and may be calculated as follows:
Let

$$\mathbf{T} = \begin{bmatrix} 2 & & \\ 1 & 3 & \\ 4 & 2 & 1 \end{bmatrix}, \qquad \text{and } \mathbf{T}^{-1} = \begin{bmatrix} a & & \\ b & c & \\ d & e & f \end{bmatrix}.$$

Then since

$$\mathbf{T}\mathbf{T}^{-1} = \mathbf{I}, \tag{3.25}$$

we have,

$$\begin{bmatrix} 2a & 0 & 0 \\ a+3b & 3c & 0 \\ 4a+2b+d & 2c+e & f \end{bmatrix} = \begin{bmatrix} 1 & 0 & 0 \\ 0 & 1 & 0 \\ 0 & 0 & 1 \end{bmatrix}$$

On equating terms in the two matrices we obtain

$$a = 1/2, \qquad c = 1/3, \qquad f = 1,$$

$$a + 3b = 0, \text{ hence } b = -1/6,$$

$$2c + e = 0, \text{ hence } e = -2/3,$$

and $$4a + 2b + d = 0, \text{ hence } d = -5/3,$$

so that

$$\mathbf{T}^{-1} = \begin{bmatrix} 1/2 & & \\ -1/6 & 1/3 & \\ -5/3 & -2/3 & 1 \end{bmatrix}.$$

For more general procedures see Lawley and Maxwell (1971, Appendix A 1.15)

## Quadratic forms and their differentiation

If $\mathbf{A}$ is a square matrix and $\mathbf{x}$ is a non-null column vector then the expression:

$$\mathbf{x}'\mathbf{A}\mathbf{x} = \sum_{i,j} a_{ij} x_i x_j \tag{3.26}$$

is called a quadratic form. If $\mathbf{x}'\mathbf{A}\mathbf{x} > 0$ the matrix $\mathbf{A}$ is said to be *positive definite:* if $>$ is replaced by $\geqslant$, $\mathbf{A}$ is said to be non-negative definite or positive semi-definite. Let us consider an example in which $\mathbf{A}$ is symmetric.

If $\mathbf{x} = \{x_1 \quad x_2\}$ and $\mathbf{A} = \begin{bmatrix} 2 & 1 \\ 1 & 3 \end{bmatrix}$,

then

$$y = \mathbf{x}'\mathbf{A}\mathbf{x} \tag{3.27}$$
$$= 2x_1^2 + 2x_1x_2 + 3x_2^2$$

The partial derivatives of $y$ with respect to $x_1$ and $x_2$ are

$$\begin{aligned} \partial y/\partial x_1 &= 4x_1 + 2x_2, \\ \partial y/\partial x_2 &= 2x_1 + 6x_2. \end{aligned} \tag{3.28}$$

In matrix notation equations (3.28) are

$$\begin{bmatrix} \partial y/\partial x_1 \\ \partial y/\partial x_2 \end{bmatrix} = 2\begin{bmatrix} 2 & 1 \\ 1 & 3 \end{bmatrix}\begin{bmatrix} x_1 \\ x_2 \end{bmatrix},$$

Hence it follows from equation (3.27) that

$$\partial y/\partial \mathbf{x} = 2\mathbf{A}\mathbf{x} \tag{3.29}$$

Equivalently, since $A$ is symmetric,

$$\partial y/\partial \mathbf{x} = 2\mathbf{x}'\mathbf{A}. \tag{3.30}$$

For **A** symmetric it can also be shown that if

$$\mathbf{y} = \mathbf{A}\mathbf{x}, \text{ then } \partial \mathbf{y}/\partial \mathbf{x} = \mathbf{A}, \tag{3.31}$$

and if

$$\mathbf{y} = \mathbf{x}'\mathbf{A}, \text{ then } \partial \mathbf{y}/\partial \mathbf{x} = \mathbf{A}. \tag{3.32}$$

The reader may also verify that if

$$y = \mathbf{x}'\mathbf{x}, \text{ then } \partial y/\partial \mathbf{x} = 2\mathbf{x}. \tag{3.33}$$

**Lagrange multiplier**

Suppose we wish to maximize $\mathbf{x}'\mathbf{A}\mathbf{x}$ subject to the condition that $\mathbf{x}'\mathbf{x} = 1$. Introduce the Lagrange multiplier $\lambda$ and set up the expression

$$K = \mathbf{x}'\mathbf{A}\mathbf{x} + \lambda(1 - \mathbf{x}'\mathbf{x}).$$

Then

$$\partial K/\partial \mathbf{x} = 2\mathbf{A}\mathbf{x} - 2\lambda\mathbf{x} \tag{3.34}$$

On equating (3.34) to zero, we obtain

$$(\mathbf{A} - \lambda\mathbf{I})\mathbf{x} = 0, \tag{3.35}$$

and the problem becomes one of finding the latent roots and vectors of **A**.

Other results required are the following. Let **A** be a square matrix of coefficients and **X** a matrix of variables (of the same order as **A**). Denote the trace of the matrix **AX** by $K$ so that

$$K = \mathrm{tr}(\mathbf{AX}). \tag{3.36}$$

Then it can be shown that:

$$\partial K/\partial \mathbf{X} = \mathbf{A}'. \tag{3.37}$$

It can also be shown that:

if $K = \mathrm{tr}(\mathbf{A}'\mathbf{X})$, then $\partial K/\partial \mathbf{X} = \mathbf{A}$; (3.38)

if $K = \mathrm{tr}(\mathbf{A}'\mathbf{X}')$, then $\partial K/\partial \mathbf{X} = \mathbf{A}'$; (3.39)

and if $K = \mathrm{tr}(\mathbf{AX}^{-1})$, then $\partial K/\partial \mathbf{X} = -(\mathbf{X}^{-1}\mathbf{AX}^{-1})'$. (3.40)

For proofs see Lawley and Maxwell (1971, Al. 16).

### Latent roots and vectors of non-symmetric matrices

In Chapters 8 and 9 we shall require a straightforward method for finding latent roots and vectors of matrices of the form $\mathbf{B}^{-1}\mathbf{A}$ which are non-symmetric despite the fact that **A** and **B** are symmetric. They arise in the solution of equations of the form

$$(\mathbf{A} - \lambda\mathbf{B})\mathbf{x} = 0, \tag{3.41}$$

which may be written, in standard form (3.11), as

$$(\mathbf{B}^{-1}\mathbf{A} - \lambda\mathbf{I})\mathbf{x} = 0. \tag{3.42}$$

The procedure is to express **B** in the form $\mathbf{B} = \mathbf{TT}'$ where **T** is a lower triangular matrix. On substitution in (3.41) we get

$$(\mathbf{A} - \lambda\mathbf{TT}')\mathbf{x} = 0.$$

Pre-multiplication within the brackets by $\mathbf{T}^{-1}$ and post-multiplication by $\mathbf{T}'^{-1}$ gives

$$(\mathbf{T}^{-1}\mathbf{AT}'^{-1} - \lambda\mathbf{I})\mathbf{y} = 0. \tag{3.43}$$

The matrix $\mathbf{T}^{-1}\mathbf{AT}'^{-1}$ is now symmetric. It has the same latent roots as $\mathbf{B}^{-1}\mathbf{A}$ but its latent vectors are altered by the transformation, the vector $\mathbf{x}$, required in (3.42), being given by

$$\mathbf{x} = \mathbf{T}'^{-1}\mathbf{y}, \tag{3.44}$$

where $\mathbf{y}$ is the corresponding latent vector of $\mathbf{T}^{-1}\mathbf{A}\mathbf{T}'^{-1}$. These results can easily be verified by pre-multiplication of (3.43) by $\mathbf{T}'^{-1}$, writing the result in the form

$$(\mathbf{T}'^{-1}\mathbf{T}^{-1}\mathbf{A} - \lambda\mathbf{I})\mathbf{T}'^{-1}\mathbf{y} = 0,$$

and comparing it with (3.42); note that $\mathbf{B}^{-1} = \mathbf{T}'^{-1}\mathbf{T}^{-1}$.

CHAPTER FOUR

# Principal Component Analysis

## Matrix transformation

In Chapter 1 a brief account was given of the origins of principal component analysis. In this chapter the technique is described in terms of matrix algebra and some of its uses discussed.

Suppose that we have scores for a sample of n individuals on each of $p$ variates, $X_j (j = 1, \ldots p)$. The scores may be written in matrix form as

$$\mathbf{X}^* = [X_{ij}],$$

where $X_{ij}$ stands for the score on the $j$th variate for the $i$th individual $(i = 1, \ldots, n)$. Denote the means of the scores by the row vector

$$\bar{\mathbf{x}} = [\bar{x}_1 \bar{x}_2 \ldots \bar{x}_p].$$

The scores in $X^*$ can be expressed as deviations from their respective means, giving the matrix

$$\mathbf{X} = [x_{ij}],$$

in which the mean of each column of scores is zero. The corrected sums of squares and cross-products of the variates is now given by $\mathbf{X}'\mathbf{X}$, where $\mathbf{X}'$ is the transpose of $\mathbf{X}$, hence the covariance matrix $\mathbf{S}$ for the variates is estimated by

$$\mathbf{S} = (\mathbf{X}'\mathbf{X})/(n-1). \tag{4.1}$$

The purpose of a principal component analysis is to transform the matrix $\mathbf{X}$ of $p$ variates, which may be correlated, into another matrix $\mathbf{Y}$ of $p$ uncorrelated hypothetical variates which decrease in variance from first to last. This is achieved by postmultiplying $\mathbf{X}$ by an orthogonal matrix $\mathbf{U}$, giving

$$\mathbf{Y} = \mathbf{X}\mathbf{U}, \tag{4.2}$$

the columns of $\mathbf{U}$ being the normalized latent vectors of the $p \times p$

matrix $\mathbf{S}$, arranged in order so that the first corresponds to the largest latent root of $\mathbf{S}$, the second to the second largest, and so on. We note that since $\mathbf{U}$ is an orthogonal matrix, $\mathbf{UU'} = \mathbf{I}$, hence

$$\mathbf{YY'} = \mathbf{XUU'X'} = \mathbf{XX'}, \tag{4.3}$$

indicating that the sum of squares of the scores for an individual is unaltered by the transformation. This result is a generalization of that given in equation (3.24) and its implications have already been noted.

Expression (4.2) may be justified in part as follows. Let the variate $y_1$, corresponding to the first column of $\mathbf{Y}$, represent the first principal component. It is a weighted sum of the variates $x_j (j = 1, \ldots, p)$ and may be written as

$$y_1 = x_1 u_{11} + x_2 u_{21} + \ldots + x_p u_{p1}, \tag{4.4}$$

in which $u_{j1}$ is the weight of the $j$th variate in this component. For the sample of $n$ individuals (4.4) can be written in matrix notation as

$$\mathbf{y} = \mathbf{Xu}, \tag{4.5}$$

in which $\mathbf{y}$ is the $n \times 1$ vector of scores on $y_1$ and $\mathbf{u}$ is the $p \times 1$ vector of weights which have to be determined.

The mean of the scores in $\mathbf{y}$ is found from (4.4) by replacing the $x$'s by their mean values, but since the latter are zero the former is also zero. The corrected sum of squares of the scores in $\mathbf{y}$ is thus given by $\mathbf{y'y}$ and their variance by $\mathbf{y'y}/(n-1)$. Using (4.5) and then (4.1) we may express the variance in the form

$$\begin{aligned} \text{var}\cdot(y_1) &= (\mathbf{u'X'Xu})/(n-1) \\ &= \mathbf{u'Su} \end{aligned} \tag{4.6}$$

The problem is to find the vector $\mathbf{u}$ which maximizes $\mathbf{u'Su}$, subject to the condition that $\mathbf{u'u} = 1$. Its solution has already been discussed in Chapter 3 (see equation 3.35) and requires us to find the largest latent root, say $\lambda_1$, and corresponding latent vector of $\mathbf{S}$.

Let $\mathbf{x}$ be a vector such that (see equation 3.10)

$$\mathbf{Sx} = \lambda_1 \mathbf{x},$$

On premultiplying on both sides by $\mathbf{x'}$ we get

$$\mathbf{x'Sx} = \lambda_1 \mathbf{x'x}. \tag{4.7}$$

Now if $\mathbf{u}$ is the normalized form of $\mathbf{x}$, then $\mathbf{u'u} = 1$ and on substituting

**u** for **x** in (4.7) we have

$$\mathbf{u}'\mathbf{S}\mathbf{u} = \lambda_1 \mathbf{u}_1' \mathbf{u}_1$$
$$= \lambda_1.$$

Hence $\lambda_1$ is the variance of $y_1$, that is of the first principal component.

To revert back from deviational scores to observed scores we replace $x_j$ by $(X_j - \bar{x}_j)$ in (4.4) and after adjustment, the equation for the first principal component becomes

$$Y_1 = y_1 + c_1 = X_1 u_{11} + X_2 u_{21} + \ldots + X_p u_{p1}, \qquad (4.4a)$$

in which $c_1 = \bar{\mathbf{x}}\mathbf{u}$.

## Appraisal of results

In practice it is frequently the case that the results of a principal component analysis are appraised in terms of the latent roots and vectors of the covariance matrix without actually calculating the component scores. To facilitate the appraisal each column vector of **U** has its elements scaled so that the sum of their squares is equal to the corresponding latent root. In other words given $\lambda_j$ and $\mathbf{u}_j$ we derive a new vector $\mathbf{w}_j$, where

$$\mathbf{w}_j = \lambda_j^{1/2} \mathbf{u}_j.$$

When the correlation matrix rather than the covariance matrix has been employed in the analysis it can be shown (Morrison, 1967, p. 226) that the elements of $\mathbf{w}_j$ may be interpreted as the correlations of the observed variates with the $j$th component, and this is helpful in interpreting the component.

## Practical applications

Principal component analysis has many practical uses. One of the most important and most straightforward of these is for data reduction. When a set of observed variates are interrelated it is frequently found that the first few components derived from them account for a large part of their variance so that, without serious loss of information, the observed variates may be replaced by a smaller set of derived variates. This may be very helpful as a prelude to further analysis of the data should this involve elaborate combinatorial procedures. For example, many cluster analysis techniques

(Everitt, 1974) are of the latter kind and often are feasible only when the number of variates is small.

Used for the purpose just described a principal component analysis is straightforward in the sense that no distributional assumptions need be made about the observed variates, nor is it necessary *en route* to try to interpret the components derived from them. In other cases an analysis may be performed as an end in itself, rather than as a means to an end, and then the experimenter generally hopes that the components he derives will reveal dimensions of variability in his data more basic than his observed variates, or more informative for descriptive purposes. The limitations of the model should now be borne in mind. For example, since it does not make specific allowance for errors of measurement in the data it should ideally be restricted to situations in which such errors are likely to be small. Moreover, since a component analysis is not invariant under changes of scale in the variates it is most appropriately used when the latter are all measured in the same metric, and have variances of similar magnitude.

*Example* 4.1

A typical set of data for component analysis are those reported by Macdonell (1902) for which the correlation matrix is given in Table 1.1. The seven latent roots of the matrix are respectively:

3.80 1.50 0.65 0.36 0.34 0.24 0.11.

The normalized vector corresponding to the first of these is:

0.276 0.212 0.295 0.438 0.456 0.450 0.436.

When it is scaled, by multiplying each of its elements by $\sqrt{3.80}$, we get the values:

0.538 0.413 0.575 0.853 0.888 0.878 0.849.

These may be interpreted as the correlations between the seven observed variates and the first component. This component accounts for a proportion $3.80/7 = 0.543$, or percentage 54.3%, of the total variance of the variates, which is very considerable. It might be described as a 'size' or, perhaps more appropriately, as a 'length of limbs' component, as it correlates most highly with the last four variates, namely *left finger length*, *forearm length*, *left foot length* and *height*.

The second component accounts for an additional 21.4% of the variance of the observed variates, and for it the scaled weights are:

$$-0.447 \quad -0.784 \quad -0.628 \quad 0.288 \quad 0.339 \quad 0.219 \quad 0.220.$$

The signs of these weights have no fundamental significance in themselves as all may be reversed without damage to the analysis, but they do indicate a classification of the variates into two subgroups, the first three variates namely *head length*, *head breadth* and *face breadth* lying in one subgroup and the remainder in the other. Hence the component distinguishes between head measurements (in particular, breadth of face) on the one hand and height and limb measurements on the other. What the component tells us is that, for people of a given 'size', there is a tendency for large head measurements to be associated with small limb measurements, and *vice versa*.

An alternative way of interpreting the information given by the first two components is seen if the variates are plotted using orthogonal axes. Relative to these axes the co-ordinates of the first, namely 'head length', are 0.538 and −0.447, and so on for the other variates. A rotation of the axes in a clockwise direction, through about 65°, now produces 'components' which correspond to the two subsets of variates already detected and could be labelled respectively a 'size of head' component and a 'length of limbs' component, though the fit could be improved if the axes were allowed to become oblique. The question of the rotation of axes is discussed more fully in the next chapter.

The third component accounts for 9.3% of the total variance and for it the scaled weights are:

$$-0.712 \quad 0.206 \quad 0.309 \quad 0.056 \quad 0.030 \quad 0.048 \quad 0.005.$$

It is concerned almost exclusively with head measurements and contrasts *head length* with *head* and *face breadth*. The remaining four components together account for only 15% of the observed variance and need not be discussed in detail here.

The example above illustrates a useful general characteristic of a component analysis, for we have seen that successive components (excluding the first, in cases where all the correlations are positive) divide the variates into subgroups which contrast with each other in some sense, and reveal relationships within the data which otherwise might go undetected. Yet, unfortunately, a classification of variates into contrasting sub-groups is only of marginal value for classifying the individuals themselves. This is so because the scores on a component, being a weighted sum of several variates, tend to be normally distributed and especially so when the variates themselves are so distributed. Hence for each component the majority of individuals have scores which cluster around the mean value, and only the relatively small numbers of individuals in the respective tails of the distribution are likely to exemplify to any marked degree the contrast which a particular classification of variates indicates. In an attempt to overcome this limitation some research workers resorted to the practice of reversing the roles of individuals and variates in an analysis, so that a classification of the former rather than the latter would be effected. Although the procedure, which is known as Q-technique, proved to be of little practical value it is worth mentioning in passing.

### Q-technique

Let $\mathbf{Z}$ be an $n \times p$ matrix $(n > p)$ in which the scores on each of the $p$ variates are in standardized form. Each variate then has zero mean and unit variance and all are expressed in 'equivalent' units. The sums of squares and cross-products for the variates is given by $\mathbf{Z'Z}$, a square symmetric matrix of order $p$ and, we will assume, of full rank. Let $\lambda$ be a latent root of this matrix and $\mathbf{y}$ be a vector such that

$$(\mathbf{Z'Z})\mathbf{y} = \lambda\mathbf{y}.$$

On premultiplying this equation on both sides by $\mathbf{Z}$ we get

$$(\mathbf{ZZ'})\mathbf{Zy} = \lambda\mathbf{Zy}. \tag{4.8}$$

Now consider the matrix $\mathbf{ZZ'}$. It is symmetric of order $n$ but of rank $p$ (since $p < n$) and so has only $p$ non-zero latent roots. It contains the sums of squares and cross-products for individuals rather than for variates. In (4.8) $\mathbf{Zy}$ is a column vector (of order $n$) and, in view of (3.10), it is the latent vector of the matrix $\mathbf{ZZ'}$ corresponding to the latent root $\lambda$. The elements of this vector may now be scaled so that their sum of squares is equal to $\lambda$ and we then have a 'component' for individuals rather than for variates. For successive components the pattern of positive and negative signs of their elements might on occasion suggest a classification of individuals were it the case that each individual had a weight on one or two components only.

The procedure has been objected to on several grounds. From a sampling viewpoint the units to be sampled become variates rather than individuals and so are unlikely to be independent of each other. In addition, a hypothetical population of variates has to be assumed and this in practice is invariably limited in size so that only very small samples are possible. From an arithmetic viewpoint it is noted that a change in the direction of scoring of even a single variate, which in itself is trivial, would alter the structure of every component. Despite these limitations some useful properties of Q-technique, in the analysis of dichotomously scored data, have been demonstrated in an interesting paper by Gower (1966).

**Extracting a 'difficulty' component**

In the last section of Chapter 2 it was noted that in the analysis of a matrix of phi-coefficients a component was likely to exist having weights closely proportional to the means of the variates. If thought desirable this 'spurious' component can be eliminated from the matrix at the outset. As before, let $p_j$ be the mean of the $j$th variate, and let $\bar{p}$ be the average mean for all variates. Construct a column vector $\mathbf{p}$ with elements $p_j - \bar{p}$. Then, if $\mathbf{R}$ represents the matrix of phi-coefficients, we want to find a constant $c$ such that the trace of the matrix $\mathbf{E}'\mathbf{E}$, where $\mathbf{E}$ is the residual matrix given by

$$\mathbf{E} = \mathbf{R} - c\mathbf{p}\mathbf{p}', \tag{4.9}$$

is minimised. The weights of the variates in the spurious component are then given by the elements of the vector $c^{1/2}\mathbf{p}$. On setting the derivative of the trace of $\mathbf{E}'\mathbf{E}$, with respect to $c$, equal to zero we obtain

$$c = (\mathbf{p}'\mathbf{R}\mathbf{p})/(\mathbf{p}'\mathbf{p})^2. \tag{4.10}$$

The analysis of the data proper is then carried out using the matrix $\mathbf{E}$, whose rank is $(p - 1)$.

CHAPTER FIVE

# Factor Analysis

## Review

The origins of factor analysis have already been noted and the suggestion made that the factor model is particularly suited to the analysis of correlational structure in the case of variates which can only be measured imprecisely, and in situations in which the structure itself may be subject to internal constraints. The model will now be described in more detail and illustrations of its use given. Over the years many different methods of estimating its parameters have been used. The most efficient of these is the maximum likelihood method and, although it involves laborious calculations, it is assumed here that computer programs are available and that it is the method being employed. It has the very liberating property that the loadings (weights) of a variate, in the factors, derived from the analysis of a correlation matrix have only to be multiplied by the standard deviation of that variate to obtain the loadings which would have been derived had the corresponding covariance matrix been analysed (see Morrison, 1967, p. 268).

Relieved of the burden of describing numerical procedures, the primary purpose of this account is to give the student some understanding of basic principles so that he may use factor analysis in an enlightened way and interpret computer output intelligently. The statistical theory is available elsewhere (e.g. Lawley and Maxwell, 1971).

## The basic model

Let $z_i (i = 1, \ldots, p)$ represent $p$ linearly independent variate which have a multivariate normal distribution in some population. Amongst other things this implies that each pair of variates has a bivariate normal distribution. Assume, for convenience, that the variates are in standardized form and denote their correlation matrix in the population by $\mathbf{\Sigma}$. This matrix is of order and rank $p$.

Now suppose that there are constrains on the correlation coefficients in $\mathbf{\Sigma}$ such that, by adjustment of its diagonal elements, its rank can be reduced from $p$ to $k$. The adjusted matrix, say $\mathbf{\Sigma}_a$, is now of rank $k$ and its $p$ rows (or columns) are no longer linearly independent. In other words, the matrix $\mathbf{\Sigma}_a$ can be taken to depend on $k$ rather than on $p$ variates $(k < p)$. In the factor model these $k$ variates, or hypothetical dimensions of variability, represent the common factors of the $p$ initial variates and are sufficient to account for their interrelationships.

The model can be expressed in algebraic terms by the linear equation

$$z_i = (\lambda_{i1} f_1 + \ldots + \lambda_{ik} f_k) + e_i, \tag{5.1}$$

which states that the $i$th variate is a weighted sum of $k$ factors $f_r (r = 1, \ldots, k)$ plus a residual variate $e_i$ peculiar to the $i$th variate itself. Hence the model postulates $(k + p)$ hypothetical variates in all, $k$ common factors and $p$ residual variates. The residual variates are taken to be uncorrelated with each other and with the factors. The latter are taken to have zero means and unit variances, that is $E(f_r) = 0$ and $E(f_r^2) = 1$. Initially too we shall assume that the factors are uncorrelated with each other, that is that $E(f_r f_s) = 0$, for $r \neq s$.

If we denote the variance of $e_i$ by $\psi_i$ then the variance of $z_i$, namely $E(z_i^2)$, can be expressed in terms of its factor loadings $\lambda_{ir}$ and its residual variance $\psi_i$. From (5.1) we have

$$\begin{aligned} E(z_i^2) &= E(\lambda_{i1} f_1 + \ldots + \lambda_{ir} f_r + e_i)^2, \\ &= \lambda_{i1}^2 E(f_1^2) + \ldots + \lambda_{ir}^2 E(f_r^2) + E(e_i^2), \\ &= \lambda_{i1}^2 + \ldots + \lambda_{ir}^2 + \psi_i. \end{aligned} \tag{5.2}$$

Since the variance of $z_i$ has been set equal to unity we obtain from (5.2) the relationship

$$1 - \psi_i = \lambda_{i1}^2 + \ldots + \lambda_{ir}^2, \tag{5.3}$$

the expressions on either side of which give the 'communality' of $z_i$, that is the part of its variance which can be attributed to the common factors. Let $\mathbf{\Psi}$ be a diagonal matrix whose $p$ diagonal elements are the residual variances $\psi_i$, then

$$\mathbf{\Sigma}_a = \mathbf{\Sigma} - \mathbf{\Psi}$$

and the diagonal elements of $\Sigma_a$ are the communalities of the variates. The correlation between any pair of the original variates, say $z_i$ and $z_j$, is $E(z_i z_j)$, and using (5.1) we find that

$$E(z_i z_j) = \lambda_{i1}\lambda_{j1} + \ldots + \lambda_{ik}\lambda_{jk}, \tag{5.4}$$

since the $f$'s and $e$'s are all uncorrelated. As an example, suppose that there are three factors and that the loadings of the two variates in them are as follows:

| | $f_1$ | $f_2$ | $f_3$ |
|---|---|---|---|
| $z_i$ | 0.80 | − 0.17 | − 0.21 |
| $z_j$ | 0.75 | − 0.30 | 0.08 |

then, using (5.4), the correlation between the variates is

$$(0.80 \times 0.75) + (-0.17 \times -0.30) + (-0.21 \times 0.08) = 0.634.$$

Also, using (5.3), the communality of $z_i$ is

$$(0.80^2) + (-0.17)^2 + (-0.21)^2$$

hence the variance $\psi_i$ of $e_i$ is $(1 - 0.713) = 0.287$, and so on.

Equation (5.1) may be expressed in matrix form as

$$\mathbf{z} = \mathbf{\Lambda f} + \mathbf{e}. \tag{5.5}$$

in which $\mathbf{z}$ and $\mathbf{e}$ are column vectors with respective elements $z_i$ and $e_i (i = 1, \ldots, p)$; $\mathbf{f}$ is a column vector with elements $f_r (r = 1, \ldots, k)$ and $\mathbf{\Lambda} = [\lambda_{ir}]$ is a $p \times k$ matrix of factor loadings. $\mathbf{\Sigma}$ is $E(zz')$ and, using (5.5), we get

$$\mathbf{\Sigma} = \mathbf{\Lambda\Lambda}' + \mathbf{\Psi}, \tag{5.6}$$

hence

$$\mathbf{\Sigma}_a = \mathbf{\Lambda\Lambda}'. \tag{5.7}$$

An overall restriction on the model is that $(p - k)^2$ be greater than $(p + k)$. If $k = 1$, $\mathbf{\Lambda}$ is a column vector of $p$ loadings and is unique apart from a possible reversal of signs of the loadings. But if $k > 1$ there is an infinity of choices of $\mathbf{\Lambda}$. To illustrate this let $\mathbf{M}$ be any orthogonal matrix of order $k$ and replace $\mathbf{\Lambda}$ by $\mathbf{\Lambda M}$ in the first term on the right hand side of (5.6). It then becomes $\mathbf{\Lambda MM'\Lambda'}$ which is equal to $\mathbf{\Lambda\Lambda}'$ since $\mathbf{MM}' = I$. Hence $\mathbf{\Sigma}$ is unchanged by the orthogonal rotations of $\mathbf{\Lambda}$ by $\mathbf{M}$. In the maximum likelihood method of analysis $\mathbf{\Lambda}$ is defined uniquely by introducing the restriction that the matrix $\mathbf{\Lambda}'\mathbf{\Psi}^{-1}\mathbf{\Lambda}$ be diagonal with elements arranged in descending order of magnitude. The advantages of doing so are explained in Lawley and Maxwell (1971, Chapter 2).

Finally, if the factors are taken to be correlated, their correlations being given by the non-singular symmetric matrix $\mathbf{\Phi}$ of order $k$, then (5.6) takes the form

$$\mathbf{\Sigma} = \mathbf{\Lambda\Phi}\ \mathbf{\Lambda}' + \mathbf{\Psi}. \tag{5.8}$$

**Using the model**

In practice the population matrix $\mathbf{\Sigma}$ is unknown and has to be replaced by an estimate of it, say $\mathbf{S}$, based on a random sample of individuals from the population. The $p$ variates are now denoted by $x_i$ rather than $z_i$ indicating that observed scores on them may be subject to errors of measurement. If the correlations in $\mathbf{S}$ are of negligible magnitude the inference is that the variates are not related and a factor analysis would serve no useful purpose. To test whether $\mathbf{\Sigma}$ does in fact differ significantly from a unit diagonal matrix we may use the criterion (Bartlett, 1950, p. 78),

$$X^2 = -\{(N-1)-(2p+5)/6\}\log_e|\mathbf{S}|, \tag{5.8}$$

in which $N$ is the sample size. For moderately large $N$, $X^2$ is distributed approximately as chi-square with $\frac{1}{2}p(p-1)$ degrees of freedom. If a significant result is obtained then the factor analysis may proceed.

At the outset it is usual to set $k = 1$, the hypothesis to be tested being that a single factor will be adequate to account for the correlations in $\mathbf{S}$. (However, if we have prior knowledge to the effect that several factors are involved we might commence by setting $k$ equal to an integer greater than unity, but always making a conservative guess). The computer program (e.g. Jöreskog, 1966) estimates the loadings of the variates in the first factor and eliminates their effect from $\mathbf{S}$. The partial correlation coefficients which remain are then tested as a whole for significance and if found to be so the value of $k$ is increased by unity. The process continues until the correlations in $\mathbf{S}$ have been accounted for at an acceptable level of significance and an appropriate value for $k$ is ascertained. Ideally this value should be checked on a replicate set of data or, if $N$ is large, the initial sample may be divided randomly into two, one for exploratory the other for confirmatory purposes.

**The residual variates**

There is some confusion in the literature about the interpretation of residual variates. The early writers, following Spearman, tended to consider them to be factors in their own right, $e_i$ being a factor specific to the variate $x_i$. Later writers, mainly for convenience in estimating $\Lambda$, treated them as random error variates. In many analyses they are undoubtedly a combination of both; yet it is well to remember that the concept of specific factors is relative rather than absolute. They will tend to be pronounced in situations in which the number of variates being analysed is relatively small and the variates themselves are a somewhat heterogeneous set, which do not correlate with each other to any great extent. But in situations in which the variates are fairly numerous (say $p > 15$) and as a whole bear a strong family relationship to each other – for example, all may be tests of cognitive ability, or rating scales of anxiety, worry and depression – the non-error variance of the variates is likely to be absorbed in the common and group factors. (Indeed, the degree to which this is so can often be checked, since independent estimates of error variance for published tests are generally available.) In such cases the variances, $\psi_i$, obtained from a factor analysis, may be taken to be estimates of 'true' error variance. The corollary of this is that the communalities of the variates are estimates of their 'true' relative variances.

*Example* 5.1

In the exploratory stages of constructing a test for assessing rigidity of behaviour in individuals in the general population a psychologist assembled nine multi-item rating scales from the literature which purported to measure, to a greater or less extent, such behaviour. Typical items from the scales were:
(a) do you like to have a place for everything and everything in its proper place?
(b) do you prefer action to planning for action?
(c) do you agree that only weaklings obey rules and regulations?
Individual items were scored on a five point scale and in what appeared to be the same direction. The items were then administered to a sample of 380 adults from the population and correlation coefficients calculated for their total scores on each of the nine multi-item scales. These are given in Table 5.1.

Table 5.1 Correlation matrix for nine rating scales ($N = 380$).

| (1) | (2) | (3) | (4) | (5) | (6) | (7) | (8) | (9) |
|---|---|---|---|---|---|---|---|---|
| 1.000 | 0.638 | 0.017 | 0.139 | 0.150 | 0.704 | 0.114 | 0.160 | 0.228 |
| | 1.000 | 0.193 | 0.097 | 0.245 | 0.570 | 0.238 | 0.196 | 0.269 |
| | | 1.000 | 0.616 | −0.181 | 0.015 | 0.543 | 0.302 | 0.041 |
| | | | 1.000 | −0.372 | 0.179 | 0.476 | 0.100 | −0.052 |
| | | | | −1.000 | 0.081 | 0.012 | 0.363 | −0.533 |
| | | | | | 1.000 | 0.102 | 0.007 | 0.172 |
| | | | | | | 1.000 | 0.213 | 0.145 |
| | | | | | | | 1.000 | 0.447 |
| | | | | | | | | 1.000 |

For an observed correlation near zero in absolute magnitude the standard error may be taken as $1/\sqrt{380} = 0.051$, and it is seen that almost a quarter of the correlations in the matrix do not reach an acceptable level of significance, while two of those that are significant have negative signs. At face value one would thus conclude that the variates do not contain a dominant common factor which might be labelled 'rigidity', and the correlational structure of the nine variates requires a more detailed analysis.

In the case of a relatively small matrix it is often possible, by visual scrutiny of its elements, to discover subsets of variates which correlate relatively highly with each other, thus forming clusters and indicating group factors in the data. For the matrix above this seems to be the case, for a close examination shows that the variates tend to fall into three clusters, as follows (1, 2, 6), (3, 4, 7) and (5, 8, 9). Once clusters have been discovered it is helpful to rearrange the variates in the matrix to correspond to them. For the nine rating scales this is done in Table 5.2 and, if we cast an eye along the diagonal of the rearranged matrix, taking the variates in groups of three, the three clusters of relatively large correlations are clearly evident. It is now obvious that the nine scales are not unidimensional in content: on the contrary they indicate three relatively distinct attitudes of mind likely to influence behaviour.

The present example was chosen for its simplicity, but experience has shown that in general it is not an easy matter to assess fully the interrelationships of a set of variates by visual inspection of their correlation matrix. Luckily the techniques of factor analysis can then be employed and the inspection carried out in a routine and

Table 5.2 Correlations for rating scales in rearranged order

| (1) | (2) | (6) | (3) | (4) | (7) | (5) | (8) | (9) |
|---|---|---|---|---|---|---|---|---|
| 1.000 | 0.638 | 0.704 | 0.017 | 0.139 | 0.114 | 0.150 | 0.160 | 0.228 |
| | 1.000 | 0.570 | 0.193 | 0.097 | 0.238 | 0.245 | 0.196 | 0.269 |
| | | 1.000 | 0.015 | 0.179 | 0.102 | 0.081 | 0.007 | 0.172 |
| | | | 1.000 | 0.616 | 0.543 | −0.181 | 0.302 | 0.041 |
| | | | | 1.000 | 0.476 | −0.372 | 0.100 | −0.052 |
| | | | | | 1.000 | 0.012 | 0.213 | 0.145 |
| | | | | | | 1.000 | 0.363 | 0.533 |
| | | | | | | | 1.000 | 0.447 |
| | | | | | | | | 1.000 |

efficient manner with the aid of computers. To illustrate this the correlation matrix already given will first be employed, but a more difficult problem is considered in Example 5.2 below.

A factor analysis was carried out on the correlations in Table 5.2 and the loadings of the nine scales on the first three factors are shown in Table 5.3, together with the residual variances, $\psi_i$. At the bottom of the table the percentage of the total variance accounted for by each factor appears, for example for the first factor this is the sum of squares of its nine loadings expressed as a percentage of 9, and so on for the other factors. Together the three factors account for

Table 5.3 Loadings of nine scales in three factors

| *Scales* | *Factor loadings* | | | *Residual variances* |
|---|---|---|---|---|
| | I | II | III | |
| 1 | 0.807 | −0.206 | −0.257 | 0.241 |
| 2 | 0.747 | −0.112 | −0.009 | 0.430 |
| 6 | 0.734 | −0.152 | −0.336 | 0.325 |
| 3 | 0.319 | 0.748 | 0.296 | 0.252 |
| 4 | 0.329 | 0.718 | −0.057 | 0.373 |
| 7 | 0.364 | 0.467 | 0.275 | 0.573 |
| 5 | 0.240 | −0.569 | 0.577 | 0.286 |
| 8 | 0.324 | 0.026 | 0.550 | 0.592 |
| 9 | 0.378 | −0.267 | 0.499 | 0.537 |
| % var. | 26.63 | 19.63 | 13.67 | 59.93 |

60% of the total variance of the scales. The estimates of the standard errors of the nine loadings in the first factor are respectively,

0.045 0.050 0.049 0.120 0.157 0.077 0.195 0.087 0.130,

and inspection shows that all the loadings (except that for scale 5) are significant. Since they are all positive we see that the investigator was not wholly in error in thinking that a general factor underlying the nine scales would be found. In addition scales 1, 2 and 6, which have high loadings on this first factor, contain items primarily of type *a* above so that it would be reasonable to label the factor *rigidity of behaviour*. Its adequacy, from a statistical viewpoint, for measuring 'rigidity' is discussed in the last section of the next chapter.

Factors II and III in Table 5.3 have their highest loadings on scales 3, 4 and 7, and 5, 8 and 9 respectively, and so reflect the remaining two clusters amongst the 9 variates, but the clustering will become more evident after the factors have been rotated.

## Factor rotation

In a factor analysis, using the maximum likelihood method, the first factor accounts for a maximum amount of the communal variance of the variates, the second for a maximum amount after the first has been removed, and so on. In this sense the factors are uniquely determined. But, as we have noted earlier, the matrix of loadings can be post-multiplied by an orthogonal matrix to produce a new matrix of loadings which is equally effective as the original set in accounting for the inter-relations of the variates. This fact is often utilized in an effort to simplify a pattern of obtained loadings, in the sense that it may be possible by rotation of $\boldsymbol{\Lambda}$ to reduce a considerable percentage of its elements to values of or near zero. It may also be possible to eliminate most of the negative signs from $\boldsymbol{\Lambda}$ and thus simplify the interpretation of the factors. In other cases the purpose of a rotation may be to derive factors which correspond to concepts which are in general use in a particular field of study, for example, the concept of 'depression' in psychiatric investigations.

In the days before computers, rotations were generally performed by subjective and graphic methods taking the factors in pairs; but more recently more objective and automatic procedures have been developed and most computer programs for factor analysis have one or other of these appended to them. The one most commonly used is the 'varimax method', which rotates the factors in such a way that the new loadings tend to be either relatively large or relatively small in absolute magnitude compared with the original ones. Despite the fact that the method is empirical and, psychologically speaking, quite 'blind', it can often effect a very useful simplification of a matrix of loadings. The varimax method retains the orthogonality between factors, but if we are prepared to relax this restriction the varimax procedure may be followed by the *promax* method (Hendrickson and White, 1964) which produces correlated factors, and this may simplify the results of an analysis even further. All these procedures are described elsewhere (Lawley and Maxwell, 1971, Chapter 6) and details are not repeated here. But it is informative to consider the results obtained in several examples in which the varimax method has been employed.

*Example* 5.1 (*continued*)

The loadings given in Table 5.3 were subjected to a varimax rotation and the outcome is reported in Table 5.4; the communalities and residual variances are unaltered by the rotation.

Comparison of the loadings in Table 5.4 with those in Table 5.3 shows that the varimax rotation reveals remarkably clearly the three clusters of variates in the data. In general such a rotation is very helpful for simplifying a large matrix of factor loadings and for

Table 5.4 Rotated loadings for the nine rating scales

| *Scales* | *Loadings* | | |
|---|---|---|---|
| | I | II | III |
| 1 | 0.863 | 0.039 | 0.116 |
| 2 | 0.683 | 0.169 | 0.274 |
| 6 | 0.821 | 0.039 | 0.004 |
| 3 | −0.016 | 0.863 | 0.062 |
| 4 | 0.151 | 0.744 | −0.228 |
| 7 | 0.096 | 0.624 | 0.169 |
| 5 | 0.092 | −0.262 | 0.798 |
| 8 | 0.043 | 0.291 | 0.567 |
| 9 | 0.179 | 0.036 | 0.656 |

bringing to light group and overlapping group factors which otherwise might go undetected. It is least effective when a very dominant general factor exists, but in such cases the general factor may be excluded from the rotation (for an example see Maxwell, 1972a, p. 9).

In Table 5.4 the cluster of rating scales indicated by factor I has already been commented on. Item 'b', given earlier, is typical of the three scales which load heavily on factor II and the scales appear to be more concerned with introversion-extraversion than with rigidity. Item 'c' above, in some measure, is typical of the cluster of scales revealed by factor III and they seem primarily to measure neuroticism but clearly are not unrelated to rigidity.

*Example* 5.2

In a study concerned with the selection of airmen, Fleishman and Hempel (1954) administered 18 tests to a sample of 197 individuals. The tests, which are listed in Table 5.5 and are numbered from (9) to (26) to agree with the numbering used by the authors, were concerned with mechanical aptitude, rate of movement, the comprehension of spatial relations, and similar problems of a technical nature. For reasons of space the correlation matrix for the tests is not reproduced here but the reader may safely assume that it presented a complicated set of interrelationships. All correlations were positive, varying in magnitude from 0.06 to 0.63. Five factors were found to be necessary and sufficient to account for them, the test of significance of the residuals giving a chi-square value of 90.83 based on 73 degrees of freedom, which is not significant.

Table 5.5 Loadings of 18 tests on first factor–Fleishman Hempel data

| *Tests* | *Loadings* | *Tests* | *Loadings* |
|---|---|---|---|
| (9) Numerical operations | 0.642 | (18) Spatial orientation | 0.689 |
| (10) Dial and table reading | 0.841 | (19) Speed of marking | 0.633 |
| (11) Mechanical principles | 0.617 | (20) Log book accuracy | 0.588 |
| (12) General mechanics | 0.481 | (21) Rotary pursuit | 0.461 |
| (13) Speed of identification | 0.760 | (22) Plane control | 0.346 |
| (14) Pattern comprehension | 0.723 | (23) Reaction time (D) | 0.678 |
| (15) Visual pursuit | 0.518 | (24) Nut and bolt | 0.369 |
| (16) Decoding | 0.742 | (25) Reaction time | 0.169 |
| (17) Instrument comprehension | 0.620 | (26) Rate of movement | 0.303 |

Table 5.6 Varimax rotated loadings for 5 factors on 18 Tests

| *Tests* | *Factors* | | | | |
|---|---|---|---|---|---|
| | I | II | III | IV | V |
| ( 9) | 0.27 | | | | 0.77 |
| (10) | 0.52 | 0.37 | | 0.32 | 0.48 |
| (18) | 0.63 | 0.25 | | 0.28 | |
| (11) | 0.39 | 0.58 | | | |
| (12) | | 0.73 | | | |
| (24) | | 0.32 | | | |
| (14) | 0.65 | 0.28 | | | |
| (15) | 0.45 | | | | |
| (16) | 0.73 | | 0.26 | | 0.27 |
| (17) | 0.54 | | | | 0.31 |
| (19) | 0.43 | | | 0.54 | |
| (13) | 0.60 | | | 0.41 | |
| (20) | | 0.28 | 0.27 | 0.66 | |
| (21) | | 0.31 | 0.45 | | |
| (22) | | | 0.48 | | |
| (23) | 0.55 | | 0.36 | | |
| (25) | | | 0.53 | | |
| (26) | | | 0.45 | | |

The first was a general factor and the loadings of the tests in it are given in Table 5.5.

The pattern of positive and negative signs of loadings on the remaining factors did not suggest a clear-cut classification of the tests and, as the general factor is clearly a combination of many facets of mechanical aptitude and co-ordination of movement, it was decided to subject all five factors to a varimax rotation in the hope that a more meaningful picture would emerge. The results are shown in Table 5.6 and, for clarity, all loadings less than 0.250 in absolute value are omitted and the order of the tests is rearranged so as to make the overlapping clusters within them more readily recognizable. A relatively clear classification of the tests now emerges and reasonable labels for the five factors are:

I Comprehension of spatial relations
II Mechanical aptitude and expertise
III Rate of movement
IV Speed and accuracy in clerical operations
V Facility with numbers.

It thus appears that the wide spectrum of information sampled by the 18 tests can be well summarized under these five labels.

## A salutary point

Only in the past decade have satisfactory numerical methods been developed for finding efficient estimates of the parameters in the factor model (see Jöreskog, 1967). When applied to existing sets of data it was found that in not a few cases one or more of the residual variances ($\psi_i$) were estimated as zero, thus suggesting that the corresponding variates were measured without error. As this is hardly plausible, the question of the extent to which data may meet the assumptions made in the model arises. To simplify the question let us suppose, (i) that in a particular instance the value of $k$, the number of common factors, has not been set at too high a level, (ii) that a check has been made on the normality of the variates, and (iii) that their sample covariance matrix $\mathbf{S}$ is positive definite. Then a likely explanation of zero $\psi$-values, if such occur, is that the assumption of independence of the residual variates is not being met and that some of the correlations are spuriously large.

To illustrate the point let us return to the correlation matrix in Table 5.2. After the three factors, whose loadings are shown in Table 5.3, have been eliminated from it, the residual partial correlation matrix was found still to be significant. An attempt was thus made to fit four factors to the data. The residual variances of the scales were now estimated to be:

| (1) | (2) | (6) | (3) | (4) | (7) | (5) | (8) | (9) |
|---|---|---|---|---|---|---|---|---|
| 0.248 | 0.393 | 0.331 | 0.020 | 0.020 | 0.626 | 0.371 | 0.577 | 0.454 |

However, two of these are altogether too small to be plausible; and one is led to suspect that errors of measurement in the data may not be independent and that it would be unwise to press for more than three factors.

Unfortunately, zero estimates of residual variances can occur early in an analysis and then they present more difficult problems of interpretation (for examples, see Maxwell, 1972a). On the computational side the usual procedure, once an offending variate or set of variates has been detected, is to eliminate their effect from the correlation matrix before the factor analysis proper begins. If a single variate only is concerned it is eliminated as a factor in its own right, in which case the factor loadings are simply the correlations of that variate with the other variates. The procedure adopted in more complicated cases is described elsewhere (e.g. Lawley and Maxwell, 1971, pp. 11–13). But the rather common occurrence of zero residuals is a salutory warning to research workers about possible limitations in their data, and it is hardly unfair to note that they are likely to go undetected if the principal component model rather than the factor model is employed.

**Factor analysis or principal component analysis?**

Despite several attempts to clearify the issue (e.g. Bartlett, 1953; Kendall and Lawley, 1956; Lawley and Maxwell, 1971, Section 1.2) a good deal of confusion still exists about the distinction between these two techniques. Basically it is this: in factor analysis, unlike principal component analysis, an hypothesis about the covariance (or correlational) structure of the variates is implied. It is formally expressed in (5.6) which states that a covariance matrix $\mathbf{\Sigma}$, of order and rank $p$, can be partitioned into two matrices $\mathbf{\Lambda\Lambda}'$ and $\mathbf{\Psi}$, the first being of order $p$ but of rank $k(k < p)$ whose off-diagonal elements are equal to the off-diagonal elements of $\mathbf{\Sigma}$; the second being a diagonal matrix of full rank $p$ whose elements when added to the diagonal elements of $\mathbf{\Lambda\Lambda}'$ give the diagonal element of $\mathbf{\Sigma}$. In other words the hypothesis is that a set $k$ of hypothetical variates $(k < p)$ exists which are adequate to account for the interrelationships of the variates though not for their full variances.

The justification for such an hypothesis in the case of psychological variates has been discussed in the fourth section of *chapter* 2, but it may also be justifiable in other fields. Whether or not a research worker wishes to consider such an hypothesis in the analysis of a set of data, and thus embark upon a factor analysis in preference to a component analysis (for which computer programs are more readily available), is likely to depend on his prior knowledge of the variates and the specific questions which he wishes to answer. The matter is taken up and discussed in more detail in the next chapter. But, in deference to the beginner, it is helpful to note that in a preliminary investigation of the correlational structure of a set of data the first few principal components, especially when they account for a fairly high percentage of the total variance of the variates are likely to provide a good overall picture. Rotation of them too by, say, the varimax method may be of additional help in detecting subgroups of variates which are especially related to each other. But he should keep in mind that if the variances of the variates are greatly inflated by error (see equation 2.5) the component weights may be spuriously large, for the first component will have extracted a maximum of the *total variance* (true and error), and succeeding components a maximum of the 'variance' which remains. It is incorrect to assume, as often seems to be the case, that error variance is reflected only in the smaller latent roots of a matrix, that is in the components which normally are discarded. Error variance is effectively eliminated only in a model which makes specific allowance for it, as is the case in the factor model.

CHAPTER SIX

# Confirmatory Factor Analysis

## The restricted factor model

The factor methods discussed in the previous chapter may be described as exploratory, for they are concerned not only with the problem of determining the number of factors required in a given case, but also with the problem of rotation which may facilitate the interpretation of factors (e.g. Example 5.2). But on occasion an investigator, experienced in some field of research, may wish to test a specific hypothesis about the factorial composition of a given set of variates in a particular population. In other words, he may feel able to postulate in advance the number of factors he expects and the pattern of zero and non-zero loadings on them. For example, he might postulate a pattern such as:

$$\begin{matrix} \times & \times & \times & \times & \times & \times & \times & \times \\ \times & \times & \times & 0 & 0 & 0 & 0 & \times, \\ 0 & \times & \times & \times & \times & \times & 0 & 0 \end{matrix}$$

involving three factors for eight variates, in which the $\times$ 's stand for non-zero loadings whose magnitudes are to be estimated. The parameters to which specific values (the zeros in this case) are assigned are referred to as *fixed* parameters, while the others are called *free* parameters. The model used (see equation 5.8) is:

$$\mathbf{\Sigma} = \mathbf{\Lambda\Phi\Lambda}' + \mathbf{\Psi}. \tag{6.1}$$

If the factors are assumed to be orthogonal then their correlation matrix $\mathbf{\Phi}$, of order $k$, will be unit diagonal and will contain $\frac{1}{2}k(k+1)$ fixed parameters. If the orthogonality restriction is relaxed, but the factors are still assumed to be unit normal variates, then the $k$ unit diagonal elements of $\mathbf{\Phi}$ are fixed parameters but its other $\frac{1}{2}k(k-1)$ distinct elements are free parameters which have to be estimated. Variations of the theme are also admisible: for example if $\mathbf{\Phi}$ is of the form

$$\begin{bmatrix} 1 & 0 & 0 \\ 0 & 1 & \times \\ 0 & \times & 1 \end{bmatrix},$$

then the first factor is postulated to be orthogonal to the second and third but the latter two are left free to correlate if by doing so a better fit to the data is achieved. In practice the $p$ elements of $\boldsymbol{\Psi}$ will usually be free parameters, and when this is so a necessary condition for all free parameters to be uniquely defined is that

$$n_1 + n_2 \geqslant k^2, \tag{6.2}$$

where $n_1$ is the number of fixed parameters in $\boldsymbol{\Lambda}$ and $n_2$ the number in $\boldsymbol{\Phi}$. In general it is not possible to lay down rules about the distribution of the zero values in $\boldsymbol{\Lambda}$; but when correlated factors are permitted each should have at least $(k-1)$ zero loadings. Given this restriction, and that implied in the inequality (6.2), a good working rule is to be conservative about the number of zero loadings postulated in the first instance. A computer program (e.g. Jöreskog and Gruvaeus, 1967) estimates the values of the free parameters in the model and does a test of goodness of fit to see if the factors account for the observed correlations between the variates. If they do, and if none of the residual variances is estimated as zero, the hypothesis is confirmed.

*Example* 6.1

To illustrate the use of the restricted factor model, data collected by Yule *et al.* (1969) on a sample of 150 children are used. They consisted of scores on each of 10 cognitive tests, from the Wechsler series, administered when the children were of average age 5 years 5 months, together with their scores on a test of reading ability administered when the children were of average age 7 years. Investigation of the data (Maxwell, 1972b) showed that when the sample was divided into two, one consisting of children above average score on the reading test (hereafter called the 'good' readers) and the other of children below average (the 'poor' readers), the structure of the correlation matrices for the 10 cognitive tests for the two subsamples differed considerably and it was decided to investigate these differences by means of factor analysis.

The names of the 10 tests and the means and standard deviations of the scores for the two subsamples are given in Table 6.1. From it we see that the good readers score higher on average than do

Table 6.1 Means and standard deviations of scores on ten cognitive tests for good and poor readers

| *Tests* | *Good readers* | | *Poor readers* | |
|---|---|---|---|---|
| | *Mean* | *S.D.* | *Mean* | *S.D.* |
| (1) Information | 11.77 | 2.63 | 9.84 | 3.01 |
| (2) Vocabulary | 12.41 | 2.56 | 9.81 | 3.17 |
| (3) Arithmetic | 11.95 | 2.55 | 9.96 | 2.39 |
| (4) Similarities | 10.76 | 2.28 | 9.52 | 2.37 |
| (5) Comprehension | 10.80 | 1.93 | 9.43 | 2.82 |
| (6) Animal house | 12.01 | 2.20 | 9.99 | 2.64 |
| (7) Picture completion | 11.17 | 2.65 | 9.52 | 2.33 |
| (8) Mazes | 12.12 | 2.45 | 10.23 | 2.95 |
| (9) Geometric design | 12.59 | 2.96 | 11.33 | 3.09 |
| (10) Block design | 12.39 | 2.25 | 10.39 | 2.44 |

Table 6.2 Correlation matrix for good readers on ten tests

| (1) | (2) | (3) | (4) | (5) | (6) | (7) | (8) | (9) | (10) |
|---|---|---|---|---|---|---|---|---|---|
| 1.000 | 0.409 | 0.332 | 0.270 | 0.483 | −0.048 | 0.091 | −0.100 | 0.137 | 0.106 |
| | 1.000 | 0.285 | 0.266 | 0.452 | 0.138 | 0.201 | −0.055 | 0.026 | 0.169 |
| | | 1.000 | 0.323 | 0.360 | 0.221 | 0.183 | 0.094 | 0.221 | 0.411 |
| | | | 1.000 | 0.262 | 0.262 | 0.164 | 0.068 | 0.222 | 0.383 |
| | | | | 1.000 | 0.201 | 0.208 | 0.165 | 0.234 | 0.252 |
| | | | | | 1.000 | 0.389 | 0.315 | 0.036 | 0.266 |
| | | | | | | 1.000 | 0.371 | 0.127 | 0.472 |
| | | | | | | | 1.000 | 0.352 | 0.432 |
| | | | | | | | | 1.000 | 0.330 |
| | | | | | | | | | 1.000 |

the poor readers, but that the dispersions of the scores in the two subsamples vary somewhat from test to test though not to any great extent.

The correlation matrix for the good readers is given in Table 6.2. With three minor exceptions, the correlations are all positive indicating the family resemblance of the tests (all cognitive) involved. But inspection of the matrix also shows that the tests tend to fall into two clusters, numbers (1) to (5) inclusive in the first cluster and the remainder in the second. This confirms prior information about the tests, for the first five are described in the literature as 'verbal' tests and the last five as 'performance' tests. The correlational structure in the matrix is also typical of that found for older children on similar tests and on the basis of the information available it was decided to test the hypothesis that three factors were involved, a general factor and two group factors, and that the pattern of zero and non-zero loadings on them would be as follows:

$$\begin{array}{cccccccccc}
\times & \times & \times & \times & \times & \times & \times & \times & \times & \times \\
0 & 0 & 0 & 0 & \times & \times & \times & \times & \times & \times . \\
\times & \times & \times & \times & \times & 0 & 0 & \times & 0 & 0
\end{array}$$

The hypothesis was confirmed, the test of goodness of fit giving a chi-square value of 20.05 based on 23 degrees of freedom which is not significant; the estimates of the non-zero loadings and of the residual variances are given in Table 6.3. But certain limitations of the data require to be noted. In the first place the sample size ($n = 75$) is relatively small and the estimates of the standard errors of the loadings are relatively large. As a consequence several of the loadings on the first two factors do not reach an acceptable level of significance and the possibility exists that a pattern of loadings involving two group factors only, might suffice to account for the intercorrelations of the tests. However, from the viewpoint of generalization of the results there is strong prior information to support the hypothesis tested. A few specific points about the results may also be noted. For example it is seen that the residual

Table 6.3 Loadings on three factors for good readers

| *Tests* | *Factors* | | | *Residual variances* |
|---|---|---|---|---|
| | I | II | III | |
| (1) Information | 0.142 | | 0.674 | 0.525 |
| (2) Vocabulary | 0.226 | | 0.571 | 0.623 |
| (3) | 0.493 | | 0.350 | 0.635 |
| (4) | 0.473 | | 0.266 | 0.706 |
| (5) | 0.258 | 0.269 | 0.658 | 0.421 |
| (6) | 0.350 | 0.229 | | 0.825 |
| (7) | 0.500 | 0.239 | | 0.693 |
| (8) | 0.331 | 0.886 | −0.257 | 0.050 |
| (9) | 0.318 | 0.294 | | 0.812 |
| (10) | 0.818 | 0.179 | | 0.299 |

Table 6.4 Correlation matrix for poor readers on ten tests

| (1) | (2) | (3) | (4) | (5) | (6) | (8) | (8) | (9) | (10) |
|---|---|---|---|---|---|---|---|---|---|
| 1.000 | 0.641 | 0.662 | 0.547 | 0.631 | 0.384 | 0.581 | 0.460 | 0.381 | 0.467 |
| | 1.000 | 0.585 | 0.547 | 0.682 | 0.240 | 0.523 | 0.351 | 0.251 | 0.554 |
| | | 1.000 | 0.428 | 0.575 | 0.336 | 0.589 | 0.500 | 0.401 | 0.536 |
| | | | 1.000 | 0.458 | 0.391 | 0.435 | 0.228 | 0.231 | 0.355 |
| | | | | 1.000 | 0.171 | 0.484 | 0.423 | 0.354 | 0.411 |
| | | | | | 1.000 | 0.412 | 0.207 | 0.100 | 0.475 |
| | | | | | | 1.000 | 0.562 | 0.228 | 0.557 |
| | | | | | | | 1.000 | 0.405 | 0.653 |
| | | | | | | | | 1.000 | 0.394 |
| | | | | | | | | | 1.000 |

variance of test (8) is estimated as 0.050 which is unrealistically small since the reliability of test results for very young children is not high. A more realistic value would be 0.300 to 0.400, and one suspects that the assumption of independent errors in the model is not being fully met. On the other hand several of the residual variances for other tests, notably numbers (4), (6), (7) and (9), are relatively large and undoubtedly contain variance due to specific factors. All the loadings on the third factor, with the exception of $-0.257$ for which the standard error is 0.140, are significant beyond the 5% level and since its five significant loadings are all on the so-called verbal tests the 'good' readers appear to have acquired a fairly well defined 'verbal' factor by the age of five and a half years.

The correlation matrix for the 'poor' readers is given in Table 6.4. It contrasts considerably with that for the 'good' readers given in Table 6.2 for, not only are the correlations as a whole much larger in magnitude, but also the cross-correlations between the two subsets of tests, (1) to (5) and (6) to (10), are now so pronounced that the existence of two group factors corresponding to them is in question. Postulation of a pattern of loadings for the 'poor' readers is thus rendered difficult. The following pattern, based partly on hindsight, was used and was found to fit the data closely (chi-square = 29.47 with 25 degrees of freedom):

$$\begin{matrix} \times & \times & \times & \times & \times & \times & \times & \times & \times & \times \\ 0 & 0 & \times & 0 & 0 & 0 & \times & \times & \times & \times \\ \times & 0 & 0 & \times & 0 & \times & \times & 0 & 0 & \times \end{matrix}$$

The estimates of the non-zero loadings are given in Table 6.5. When compared with their standard errors all the loadings, with the exception of 0.178 on factor II and 0.143 on factor III, were found to be significant. The dominant features of the analysis are the very pronounced general factor and the absence of a verbal group factor; these two features bring out clearly the contrast in the correlational structure of the tests for the two groups of children.

It is informative too (Maxwell *et al.*, 1974) to compare the results given in Tables 6.3 and 6.5 in terms of Thomson's model of the brain, mentioned in Chapter 1. On the basis of it a group factor indicates that the tests which contribute to it sample selectively from the basic elements (neurons) of the brain rather than from the elements as a whole. Despite this, the proportion of all the elements sampled by each test (see Maxwell 1972a) is given by the square of the loading of that test on the general factor. From our results it is clear that the 'good' readers were capable of classifying the tests into their known 'verbal' and 'performance' subsets and, apparently as a consequence of this, were able to answer the tests by employing relatively small proportions (on average) of basic mental elements, for the loadings of the tests on the general factor in Table 6.3 are relatively small. However, this is not the full story for (as noted earlier) there is evidence of specific factors in the analysis of the data for 'good' readers. This suggests that these children were mentally aware of special characteristics of different tests, in addition

Table 6.5 Loadings on three factors for poor readers

| *Tests* | *Factors* | | | *Residual variances* |
|---|---|---|---|---|
| | I | II | III | |
| (1) Information | 0.814 | | 0.143 | 0.317 |
| (2) Vocabulary | 0.824 | | | 0.321 |
| (3) Arithmetic | 0.748 | 0.178 | | 0.409 |
| (4) Similarities | 0.616 | | 0.218 | 0.573 |
| (5) Comprehension | 0.790 | | | 0.376 |
| (6) Animal house | 0.307 | | 0.925 | 0.050 |
| (7) Picture completion | 0.648 | 0.286 | 0.210 | 0.446 |
| (8) Mazes | 0.490 | 0.785 | | 0.144 |
| (9) Geometric design | 0.402 | 0.269 | | 0.766 |
| (10) Block design | 0.570 | 0.456 | 0.290 | 0.366 |

to the resemblances which enabled them to be classified into two subsets. Indeed both problems arise in any classification procedure.

When we turn to the results for the 'poor' readers a contrasting picture is seen. They obviously had not yet acquired the mental facility to identify the first five, or 'verbal' tests, as a subset since for them a verbal group factor does not emerge. In answering these tests the 'poor' readers appear, in terms of Thomson's model, to have sampled extensively and in an unselected fashion from the complete pool of mental elements, as the high loadings on factor I in Table 6.5 indicate. But this proved to be a somewhat uneconomic procedure since their mean scores on the tests are lower (Table 6.1) than those for the 'good' readers. A general deduction seems to be that ability to classify material is an important feature of mental efficiency. A more specific and indeed salutary point, suggested by the analyses, is that 'neural connections and patterns' which facilitate the process of learning to read appear to be determined at a relatively early age, that is well before reading itself is mastered.

## ESTIMATING FACTOR SCORES

The estimation of the scores of individuals on a factor presents special problems since in the model discussed above and that in Chapter 5 the number of variates postulated, namely $(p+k)$, exceeds the number of observed variates, $p$. Two methods, which are described in detail by Lawley and Maxwell (1971, Chapter 8), are briefly considered here.

The first is the regression method given originally by Thomson. To introduce it we refer back to equation (5.5) and recall the assumptions made. In particular we note that since the variates $z_i$, namely the elements of $\mathbf{z}$, and the factors $f_r$ which are the elements of $\mathbf{f}$, have unit variances the correlations between the variates and the factors are given by $E(\mathbf{zf}')$. Using (5.5) we thus have

$$\begin{aligned} E(\mathbf{zf}') &= E\{(\mathbf{\Lambda f}+\mathbf{e})\mathbf{f}'\} \\ &= \mathbf{\Lambda} E(\mathbf{ff}') \qquad (6.3) \\ &= \mathbf{\Lambda}, \end{aligned}$$

since $E(\mathbf{ff}')=\mathbf{I}$ and $E(\mathbf{ef}')=0$. The multiple regression equation (see Chapter 7) for estimating the factors is thus given by

$$\hat{\mathbf{f}} = \mathbf{\Lambda}'\mathbf{\Sigma}^{-1}\mathbf{z},$$

which can be expressed in the simpler form

$$\hat{\mathbf{f}} = (\mathbf{I} + \mathbf{\Gamma})^{-1}\mathbf{\Lambda}'\mathbf{\Psi}^{-1}\mathbf{z}, \tag{6.4}$$

in which $\mathbf{\Gamma} = \mathbf{\Lambda}'\mathbf{\Psi}^{-1}\mathbf{\Lambda}$ is a diagonal matrix of order $k$, and $\mathbf{\Psi}$ is also diagonal.

The errors of prediction of the factors are the elements of $(\hat{\mathbf{f}} - \mathbf{f})$, and their covariance matrix is

$$E[(\hat{\mathbf{f}} - \mathbf{f})(\hat{\mathbf{f}} - \mathbf{f})'] = (\mathbf{I} + \mathbf{\Gamma})^{-1} \tag{6.5}$$

For good prediction we require the diagonal elements of $(\mathbf{I} + \mathbf{\Gamma})^{-1}$ to be small, that is the diagonal elements of $\mathbf{\Gamma} = \mathbf{\Lambda}'\mathbf{\Psi}^{-1}\mathbf{\Lambda}$ to be large.

The estimates of the factor scores found by (6.4) have the disadvantage that they are biased in the sense that their expected mean value is not equal to the mean of the true factor scores. But estimates of factor scores which are not biased in this sense can be found by an alternative procedure due to Bartlett. In it the principle adopted is that of minimizing the sum of squares of the standardised residuals, which is

$$\begin{aligned}\Sigma(e_i^2/\psi_i) &= \mathbf{e}'\mathbf{\Psi}^{-1}\mathbf{e}\\ &= (\mathbf{z} - \mathbf{\Lambda}\mathbf{f})'\mathbf{\Psi}^{-1}(\mathbf{z} - \mathbf{\Lambda}\mathbf{f}).\end{aligned} \tag{6.6}$$

The estimation equation for factor scores in this case is found to be

$$\hat{\mathbf{f}}^* = \mathbf{\Gamma}^{-1}\mathbf{\Lambda}'\mathbf{\Psi}^{-1}\mathbf{z} \tag{6.7}$$

and, corresponding to (6.5), we have

$$E[(\hat{\mathbf{f}}^* - \mathbf{f})(\hat{\mathbf{f}}^* - \mathbf{f})'] = \mathbf{\Gamma}^{-1} \tag{6.8}$$

The two sets of estimates are simply related, and in the case of orthogonal factors the relationship is

$$\hat{\mathbf{f}}^* = (\mathbf{I} + \mathbf{\Gamma}^{-1})\hat{\mathbf{f}}, \tag{6.9}$$

which, given $\hat{\mathbf{f}}$ or $\hat{\mathbf{f}}^*$, involves only scaling constants, when $\mathbf{\Gamma}$ is diagonal.

As an illustration, an equation is now found for estimating individuals' scores on the 'rigidity' factor of Example 5.1. In this case $\mathbf{\Lambda}$ is the matrix of loading in Table 5.3 and $\mathbf{\Psi}$ is the diagonal matrix with the residual variances in that table as elements. We find

$$\Gamma = \Lambda'\Psi^{-1}\Lambda = \begin{bmatrix} 7.238 & -0.002 & 0.000 \\ -0.002 & 5.534 & 0.002 \\ 0.000 & 0.002 & 3.252 \end{bmatrix},$$

which is virtually diagonal, as postulated.

If we elect for unbiased estimates we use expression (6.7). On doing the calculations we find the first row of the matrix to be

$$[0.464 \quad 0.240 \quad 0.312 \quad 0.176 \quad 0.122 \quad 0.088 \quad 0.116 \quad 0.076 \quad 0.097],$$

and the estimation equation required is

$$\hat{\mathbf{f}}_1^* = 0.464z_1 + 0.240z_2 + 0.312z_6 + \ldots + 0.097z_9,$$

in which the $z$'s are standardized scores on the rating scales.

To see how adequately the factor is estimated we use expression (6.8) and find the covariance matrix for the errors of prediction. It is

$$\begin{bmatrix} 0.138 & 0.000 & 0.000 \\ 0.000 & 0.181 & 0.000 \\ 0.000 & 0.000 & 0.308 \end{bmatrix},$$

and since its first element 0.138, which refers to the factor in question, is fairly near zero we conclude that this factor is reasonably well predicted. The relative weights of the scales in the equation also suggest that prediction could be improved if further scales, similar in content to numbers 1, 2 and 6, were included in the battery. It appears too from the results shown in Table 5.4 that such a procedure would be likely to improve the validity of the factor since the remaining scales define two additional factors relatively distinct from 'rigidity'.

**Oblique factors**

The factors, whose loadings are shown in Table 5.4, were restricted to being orthogonal and as a final example of the use of the restricted model it was decided to reanalyse the correlation matrix (Table 5.2) concerned. Three correlated factors were now postulated and the pattern of zero and non-zero loadings employed for them was:

$$\begin{matrix} \times & \times & \times & 0 & \times & 0 & 0 & 0 & \times \\ 0 & \times & 0 & \times & \times & \times & \times & \times & 0 \\ \times & \times & 0 & 0 & \times & \times & \times & \times & \times \end{matrix}.$$

Table 6.6 Estimates of loadings for three correlated factors

| *Scales* | *Factors* | | |
|---|---|---|---|
| | I | II | III |
| (1) | 0.854 | | 0.110 |
| (2) | 0.653 | 0.127 | 0.263 |
| (6) | 0.819 | | |
| (3) | | 0.859 | |
| (4) | 0.175 | 0.754 | −0.279 |
| (7) | | 0.623 | 0.144 |
| (5) | | −0.302 | 0.830 |
| (8) | | 0.265 | 0.546 |
| (9) | 0.104 | | 0.660 |
| | Correlations | between | Factors |
| | 1.000 | 0.052 | 0.122 |
| | | 1.000 | 0.144 |
| | | | 1.000 |

The estimates of the non-zero loadings and of the factor correlations are given in Table 6.6.

When compared with their standard errors all the factor loadings were found to be significant, but clearly a pattern which did not yield negative loadings, could such be found, would be more satisfactory than that used above. The correlations between the factors are of negligible magnitude and it appears that in this instance relaxation of the orthogonality restriction between factors has little effect on the results.

CHAPTER SEVEN

# Multiple Linear Regression

## Introduction

The multiple linear regression model is widely used in the social sciences for prediction purposes. One of its earliest applications was in the assessment of selection procedures for assigning children to different types of secondary school, but it has also been used as an aid to personnel selection in industry, in investigations of remedial treatment of recidivists, and in many similar fields of enquiry. To use the model we require scores for a random sample of individuals from the population of interest on a number of predictor variates, the so-called *independent* variates, and also scores on a criterion variate, the so-called *dependent* variate, which we wish to predict. The problem is to find a set of weights to apply to the independent variates which will maximize the correlation between their combined effect on the one hand and the dependent variate on the other. The weights are known as *partial regression coefficients* and the correlation coefficient as the *multiple correlation coefficient*. The basic assumption made is that the variates (including the dependent variate) have a multivariate normal distribution so that all relationships between them may be taken to be linear.

Yet this assumption is not what is ordinarily made when the model and its sampling theory are discussed. Instead the independent variates are taken to be non-random, say variates which are under the control of the experimenter. Only the dependent variate is random and it is sampled at specific levels of, or points on, the independent variates. Under these and other conditions it is possible readily to find unbiased estimators for the regression coefficients in a linear model and estimates of sampling errors. But the assumption of multivariate normality is the only reasonable one which a social scientist can make in view of the nature of his data, and he has to live with the fact that any use he may make of sampling theory, derived for the more precise (or what we shall in future refer to as the classic) model, is not fully justified. Under his assumption too the estimates of the regression coefficients will be biased when the dependent variates are subject to measurement error, but a method for dealing with the problem is discussed in the last section of this chapter and an illustrative example given.

## The model

Let $X_j (j = 1, \ldots, p)$ represent the $p$ independent variates and $Y$ represent the dependent variate. For convenience we shall assume initially that the variates are all measured about their mean values and shall represent them by the lower case letters $y, x_1, x_2$, etc. Consider the linear equation

$$y = b_1 x_1 + b_2 x_2 + \ldots + b_p x_p + e, \qquad (7.1)$$

in which the $b$'s are weights. It expresses $y$ as a weighted sum of the $x$'s, together with a residual variate $e$. Since the means of $y$ and the $x$'s are zero the mean of $e$ is also zero.

Now suppose that we have scores for a random sample of size $n$ from a population on each of the $(p + 1)$ variates. The scores on $y$ can be represented by a column vector of order $n$ and those on the $x$'s by an $n \times p$ matrix $\mathbf{X}$. For the $n$ individuals equation (7.1) may then be written in matrix form as

$$\mathbf{y} = \mathbf{X}\mathbf{b} + \mathbf{e}, \qquad (7.2)$$

in which $\mathbf{b}$ is a column vector of order $p$ whose elements are the weights in (7.1), and $\mathbf{e}$ is a column vector of order $n$ whose elements are the residual scores, one for each member of the sample. We shall assume that these elements are independent of each other and are approximately normally distributed about their mean of zero. Equation (7.2) may be written as

$$\mathbf{e} = \mathbf{y} - \mathbf{X}\mathbf{b}, \qquad (7.3)$$

and it is clear that for optimal prediction of the $y$'s the $b$'s should be chosen so as to minimize the $e$'s. Since the mean of the latter is zero one way of achieving this is to minimize $\mathbf{e}'\mathbf{e}$, namely the sum of squares of the residuals. Using (7.3) we have

$$\begin{aligned} \mathbf{e}'\mathbf{e} &= (\mathbf{y} - \mathbf{X}\mathbf{b})'(\mathbf{y} - \mathbf{X}\mathbf{b}) \\ &= \mathbf{y}'\mathbf{y} - \mathbf{y}'\mathbf{X}\mathbf{b} - \mathbf{b}'\mathbf{X}'\mathbf{y} + \mathbf{b}'\mathbf{X}'\mathbf{X}\mathbf{b} \\ &= \mathbf{y}'\mathbf{y} - 2\mathbf{y}'\mathbf{X}\mathbf{b} + \mathbf{b}'\mathbf{X}'\mathbf{X}\mathbf{b}, \end{aligned} \qquad (7.4)$$

since $\mathbf{y}'\mathbf{X}\mathbf{b} = \mathbf{b}'\mathbf{X}'\mathbf{y}$, each being the same scalar quantity. We now differentiate (7.4) with respect to $\mathbf{b}$ and set the result equal to zero. This gives

$$-2\mathbf{y}'\mathbf{X} + 2\mathbf{b}'(\mathbf{X}'\mathbf{X}) = \mathbf{0}$$

or

$$\mathbf{b} = (\mathbf{X}'\mathbf{X})^{-1}\mathbf{X}'\mathbf{y}, \tag{7.5}$$

and we have an expression for determining the $b$-values which minimise $\mathbf{e}'\mathbf{e}$. In (7.5) it is assumed that the inverse of $\mathbf{X}'\mathbf{X}$ exists, but this will be so if the variates are linearly independent and $n > p$. Having found the $b$-values, or partial regression coefficients, the prediction equation is

$$\hat{y} = b_1x_1 + b_2x_2 + \ldots + b_px_p, \tag{7.6}$$

where $\hat{y}$ is the estimate of $y$. Expressed in raw scores units (7.6) becomes

$$\hat{Y} = a + b_1X_1 + b_2X_2 + \ldots + b_pX_p, \tag{7.7}$$

where $a$ is a constant with the value

$$a = \bar{Y} - (b_1\bar{X}_1 + \ldots + b_p\bar{X}_p),$$

in which $\bar{Y}$ and the $\bar{X}$'s are the sample mean values of the variates.

In matrix notation the prediction equation is

$$\hat{\mathbf{y}} = \mathbf{X}\mathbf{b}, \tag{7.8}$$

and the elements of $\hat{\mathbf{y}}$ have zero mean. The sum of squares of the $y$-scores is $\mathbf{y}'\mathbf{y}$ and that part of it which can be accounted for by the predicted scores is $\hat{\mathbf{y}}'\hat{\mathbf{y}}$. Using (7.8) we have

$$\hat{\mathbf{y}}'\hat{\mathbf{y}} = \mathbf{b}'\mathbf{X}'\mathbf{X}\mathbf{b}, \tag{7.9}$$

and the sum of squares of the residual terms is then given by

$$\mathbf{e}'\mathbf{e} = \mathbf{y}'\mathbf{y} - \mathbf{b}'\mathbf{X}'\mathbf{X}\mathbf{b}. \tag{7.10}$$

These results may conveniently be expressed in analysis of variance form, and we have

| *Source of variation* | *d.f.* | *Sums of squares* |
|---|---|---|
| Due to regression | $p$ | $\mathbf{b}'\mathbf{X}'\mathbf{X}\mathbf{b}$ |
| Residual | $n-p-1$ | $\mathbf{y}'\mathbf{y} - \mathbf{b}'\mathbf{X}'\mathbf{X}\mathbf{b}$ |
| Total | $n-1$ | $\mathbf{y}'\mathbf{y}$ |

From the analysis it follows that the variance, $s^2$, of the residual terms is given by

$$s^2 = (\mathbf{y}'\mathbf{y} - \mathbf{b}'\mathbf{X}'\mathbf{X}\mathbf{b})/(n - p - 1). \tag{7.11}$$

The multiple correlation coefficient, say $r_{\mathrm{m}}$, is the correlation between the elements of $\mathbf{y}$ and $\hat{\mathbf{y}}$. In terms of the vectors, it is given by

$$r_{\mathrm{m}}^2 = (\mathbf{y}'\hat{\mathbf{y}})^2/(\mathbf{y}'\mathbf{y}\cdot\hat{\mathbf{y}}'\hat{\mathbf{y}}). \tag{7.12}$$

But $\mathbf{y}' = \hat{\mathbf{y}}' + \mathbf{e}'$, hence $\mathbf{y}'\hat{\mathbf{y}} = (\hat{\mathbf{y}}' + \mathbf{e}')\hat{\mathbf{y}} = \hat{\mathbf{y}}'\hat{\mathbf{y}}$, since the residual and estimated scores are independent. On substituting the last result in (7.12), and then using (7.9), we get

$$r_{\mathrm{m}}^2 = (\hat{\mathbf{y}}'\hat{\mathbf{y}})/(\mathbf{y}'\mathbf{y}) = (\mathbf{b}'\mathbf{X}'\mathbf{X}\mathbf{b})/(\mathbf{y}'\mathbf{y}), \tag{7.13}$$

that is the square of the multiple correlation coefficient is the ratio of the sum of squares due to prediction to the total sum of squares. By finding the mean squares and their ratio in the analysis of variance table we obtain an $F$-ratio test for assessing the significance of $r_{\mathrm{m}}$. It can be written in the form

$$F = \{(n - p - 1)/p\}\{r_{\mathrm{m}}^2/(1 - r_{\mathrm{m}}^2)\}, \tag{7.14}$$

with $p$ and $(n - p - 1)$ degrees of freedom.

## Scaling

In the social sciences the variates employed in a regression analysis generally have arbitrary scales and it is usual to employ standardized scores rather than raw scores in the calculations. By a little algebra it can be shown that the vector of regression weights is then given by

$$\mathbf{b}^* = \mathbf{S}^{-1}\mathbf{c}, \tag{7.15}$$

in which $\mathbf{S}$ is the correlation matrix for the independent variates and $\mathbf{c}$ is the column vector of correlations between these and the dependent variate. The $j$th element $b_j$ of $\mathbf{b}$ (7.5) is then related to the corresponding element $b_j^*$ of $\mathbf{b}^*$ by the equation

$$b_j = (s_y b_j^*)/s_j, \tag{7.16}$$

in which $s_y$ is the standard deviation of $Y$ and $s_j$ that of $X_j$. The multiple correlation coefficient, $r_{\mathrm{m}}^*$, is given by

$$r_{\mathrm{m}}^{*2} = \mathbf{c}'\mathbf{S}^{-1}\mathbf{c}, \tag{7.17}$$

and its significance may be tested by substituting $r_{\mathrm{m}}^{*2}$ for $r_{\mathrm{m}}^2$ in (7.14).

## Precautions

Whereas the arithmetic calculations involved in a multiple regression analysis can be carried out, and indeed frequently are, on any available set of data, confidence in the results obtained will be increased if certain elementary precautions are taken. For instance the initial assumption of multivariate normality implies, amongst other things, that each of the $(p + 1)$ variates has a normal distribution. This can readily be checked in advance and some function of an observed variate, such as its logarithm or square root, may prove preferable to the variate itself to achieve normality. The residual variate values can also be calculated and the assumption that they are normally distributed can be checked. This assumption is basic for the validity of the test of significance of the multiple correlation coefficient (7.14).

In the case of the classic model estimates of the error variances of the partial regression coefficients are given by the diagonal elements of $(\mathbf{X}'\mathbf{X})^{-1}s^2$, where $s^2$ is found by (7.11). For it the independent variates are non-random and $\mathbf{X}$ is taken to be a matrix whose elements are constants. But if in fact the independent variates are random variates, and especially if they are subject to measurement error, then the elements of $\mathbf{X}$ themselves will be subject to variation and this will not be taken into account when the simple expression $(\mathbf{X}'\mathbf{X})^{-1}s^2$ is used to estimate the error variances. The point must be kept in mind when interpreting computer output, for the program used is likely to provide tests of significance of the regression coefficients irrespective of the limitations of the data, and these may not be reliable.

## Selective sampling

In a prediction study it is frequently the case that the data available for setting up a regression equation refer to a highly selected sample from a population rather than to one which might be considered representative of it. For example, 100 school leavers in the vicinity of a given factory might apply to be trained as technicians in it. Assume that only 40 vacancies exist and that the selection procedure consists of several aptitude tests and personality assessments, 40 applicants being chosen as a result of their scores on these. Should the factory management subsequently wish to assess the validity of its selection procedure, using as a criterion a rating of the proficiency of the new trainees, it would of necessity have to confine its calculations to the data for the 40 successful applicants. But since they are a highly selected sample from all the applicants for the vacancies the correlation coefficients between the predictor variates and the criterion are likely to be much smaller than they would have been had all 100 applicants been admitted for training. As a consequence the multiple correlation coefficient obtained, if the regression model is employed, will give an underestimate of the true value of the selection procedure, and a correction for selection is desirable.

The effect of selective sampling on the means, variances and covariances of a multivariate population was first investigated by Karl Pearson who supplied formulae to correct for it. These formulae were later expressed in matrix notation by Aitken (1934). At first it was widely believed that they were valid only in situations in which the variates were normally distributed both before and after selection, but it is now known that they are true under much more general conditions (Lawley, 1943). The problem will not be discussed in detail here though a few formulae which are of immediate interest (see Example 7.1) are supplied.

Let the square symmetric matrices **U** and **V** represent the covariances of $p$ variates in a selective sample and in the parent population respectively. Let the variates be divided into two sets and partition the matrices to correspond to them so that

$$\mathbf{U} = \left[\begin{array}{c|c} \mathbf{U}_{11} & \mathbf{U}_{12} \\ \hline \mathbf{U}_{21} & \mathbf{U}_{22} \end{array}\right] \quad \text{and} \quad \mathbf{V} = \left[\begin{array}{c|c} \mathbf{V}_{11} & \mathbf{V}_{12} \\ \hline \mathbf{V}_{21} & \mathbf{V}_{22} \end{array}\right].$$

Suppose that all the submatrices in **U** are known but that only the submatrix $\mathbf{V}_{22}$ in **V** is known and that we wish to estimate $\mathbf{V}_{11}$ and $\mathbf{V}_{21}$, $\mathbf{V}_{12}$ being simply the transpose of $\mathbf{V}_{21}$. Then by formulae supplied by Karl Pearson we have, in matrix notation,

$$\mathbf{V}_{21} = \mathbf{V}_{22}\mathbf{C},$$

and

$$\mathbf{V}_{11} = \mathbf{D} + \mathbf{C}'\mathbf{V}_{22}\mathbf{C},$$

in which $\mathbf{C} = \mathbf{U}_{22}^{-1}\mathbf{U}_{21}$ and $\mathbf{D} = \mathbf{U}_{11} - \mathbf{U}_{12}\mathbf{C}$. In the hypothetical example given above $\mathbf{U}_{11}$ would be a scalar quantity representing the variance of the criterion variate for the 40 successful candidates, $\mathbf{U}_{12}$ would be the vector of covariances of the criterion and the predictor variates, while $\mathbf{U}_{22}$ would be the covariance matrix for the latter. On the other hand $\mathbf{V}_{22}$ would be the covariance matrix for all 100 applicants (the population) on the predictor variates. The estimated matrix $\mathbf{V}$, rather than $\mathbf{U}$, would now be employed in setting up the regression equation.

*Example* 7.1

Emmett (1942) used the multiple regression model in an attempt to assess the efficiency of procedures for selecting pupils for secondary education. His independent variates were

$$x_1 = \text{an intelligence test,}$$

$$x_2 = \text{an attainment test in English,}$$

$$x_3 = \text{an attainment test in Arithmetic,}$$

administered to a population of children in the age range 10–11 years while they were still at primary schools. For a sample of 765 children who obtained secondary school places he also had teachers' assessments of their progress three years later and these constituted his dependent variate $y$. The intercorrelations of the four variates for the sample of 765 children were found to be

| $y$ | $x_1$ | $x_2$ | $x_3$ |
|---|---|---|---|
| 1.000 | 0.454 | 0.355 | 0.330 |
| 0.454 | 1.000 | 0.437 | 0.357 |
| 0.335 | 0.437 | 1.000 | 0.194 |
| 0.330 | 0.357 | 0.194 | 1.000 |

The partial regression coefficients for the independent variates were found (7.15) to be

$$0.318 \qquad 0.160 \qquad 0.185,$$

respectively, and the multiple correlation coefficient to be 0.509, all being highly significant. Now $0.509^2 = 0.259$ is the proportion of the variance of the dependent variate which can be predicted from the three independent variates and at face value one would conclude that prediction was poor. However, Emmett pointed out that these results did not give a true picture of the efficiency of the selection procedure since they are based on a highly selected sample, namely the 765 children from the whole population of primary school children whose scores on $x_1, x_2$ and $x_3$ were sufficiently high to secure for them secondary school places.

To amend his results he noted the effect of selection on the standard deviations of the raw scores of the variates. For the sample and the population they were respectively:

| *Variate* | *Sample* | *Population* |
|---|---|---|
| $y$ | $s_y = 11.23$ | $\sigma_y$ (unknown) |
| $x_1$ | $s_1 = 9.67$ | $\sigma_1 = 15.00$ |
| $x_2$ | $s_2 = 12.77$ | $\sigma_2 = 23.44$ |
| $x_3$ | $s_3 = 16.01$ | $\sigma_4 = 25.89$ |

The population standard deviation, $\sigma_y$, was unknown since teachers' assessments of children who had not obtained secondary school places naturally were not available. But using Karl Pearson's selection equations applied to the case in which one variate, $y$, is influenced by induced selection through three other variates, $x_1, x_2$ and $x_3$, he estimated it to be 14.97. As a by-product of his calculations he was also able to estimate the population correlations for $y$ with the $x$'s, the other population correlations being known; the complete matrix was:

| $y$ | $x_1$ | $x_2$ | $x_3$ |
|---|---|---|---|
| 1.000 | 0.736 | 0.702 | 0.694 |
| 0.736 | 1.000 | 0.833 | 0.809 |
| 0.702 | 0.833 | 1.000 | 0.769 |
| 0.694 | 0.809 | 0.769 | 1.000 |

and from it the population partial regression coefficients for the independent variates were found to be

0.370 0.220 0.225,

respectively, and the multiple correlation with the dependent variate to be 0.764.

Although the multiple correlation still falls far short of perfection,

Emmett comments that it is about as high as could be expected in view of the known unreliability of teachers' assessments. The best single predictor is $x_1$, the intelligence test, and since the estimate of its correlation with $y$ in the population is 0.736 it is almost as good on its own for prediction purposes as the best weighted function of all three independent variates. This is due to the fact that these variates are highly correlated. Emmett also notes the interesting theoretical point that had the degree of selection, as measured by $s_j/\sigma_j$, been the same in each independent variate, the *relative* values of the regression coefficients for standardized scores would have been the same in both sample and population.

### Selecting independent variates

It is intuitively obvious that in trying to get a high multiple correlation coefficient we should select predictor variates which correlate minimally with each other but markedly with the dependent variate. Under these conditions it is also to be expected that $r_m$ will increase somewhat as the number of predictors is increased. Less obvious is the fact that $r_m$ may be increased by including one or more variates which, though they correlate negligibly with the dependent variate, correlate significantly with the other predictor variates. They are known as *suppressor* variates and their role in prediction is well discussed by Lubin (1957). As a simple example consider the correlation matrix:

$$\left[\begin{array}{c|cc} 1 & r_{12} & 0 \\ \hline r_{21} & 1 & r_{23} \\ 0 & r_{32} & 1 \end{array}\right],$$

in which the first variate is the dependent variate ($y$) and the other two are independent variates, the second of which has zero correlation with $y$. By (7.17) we find that

$$r_m^2 = r_{12}^2/(1 - r_{23}^2).$$

The denominator in the latter expression is less than unity, hence $r_m$ is greater in absolute value than $r_{12}$, and this is so irrespective of the signs of $r_{12}$ and $r_{13}$. It follows that despite the fact that the second independent variate has zero correlation with $y$ its inclusion in the regression equation is beneficial.

Because of the complicated interrelationships which may exist between the set of variates it is difficult to select by inspection a 'good' set for predicting a criterion variate. Experience too has shown that a few of the variates available may well be as effective as the complete set for the purpose. Since the advent of computers, procedures have been developed to assist in making a choice. One of these, described by Efroymson (1965), will serve as an illustration. In the absence of other information it commences by including in the regression equation the single independent variate having the highest absolute correlation with the dependent variate. The residual sum of squares when it is employed is then calculated, as are the partial covariances of the remaining independent variates with the dependent variate. The latter are then employed to find the next best variate to include in the equation, that is the variate which most reduces the residual sum of squares. The method proceeds in this stepwise manner adding variates one at a time to the equation, but as it does so continuous checks are made to ensure that variates already in the equation merit retention as new variates are added. The process ends after all the available independent variates have been checked and variates dropped at one stage of the analysis have been rechecked for possible re-entry into the equation. The decision to retain a variate is made by a statistical test which checks on whether or not it significantly reduces the residual sum of squares, but initially the test is set at a lenient level to ensure that a good combination of several variates is not overlooked.

## Appraisal

In general the use of the multiple linear regression model for prediction in the social sciences has proved disappointing. Not only have the multiple correlation coefficients obtained been small in magnitude but also the partial regression weights have been found to be unstable in replicated studies. A typical example is one reported in an interesting study by Simon (1971). Her chief aim was to devise an instrument, based on pre-probation data, which would separate young men placed on probation into likely successes and failures, thus facilitating an evaluation of probation treatment. Data for a sample of 539 men were available, the independent variates consisting of information provided by probation officers on variates such as family background, employment, interests, and so forth. The criterion to be predicted was dichotomous in form, those persons reconvicted at least once within a year after the earlier conviction being defined as failures. The sample was divided randomly into two subsamples of sizes 270 and 269 respectively. The first was used to construct the regression equation and the second to check its validity. A stepwise analysis was used and of 15 independent variates initially employed 5 were retained in the final equation, and these gave a multiple correlation coefficient of 0.41 (Simon, 1971, p. 80). The regression weights obtained were now used, in a cross-validation study, as weights for the data in the second sample, but the multiple correlation coefficient was found to be only 0.16. In other words the set of weights which were optimal for the first sample were seen not to have good general validity.

One of the reasons for disappointing results undoubtedly lies in the unreliability of the variates with which social scientists have to deal. We recall (Example 7.1) that Emmett on obtaining a multiple correlation of 0.764, in his assessment of selection procedures for secondary education, comments that this value is about as high as could be expected in view of the limited reliability of his dependent variate. His statement is based on a well-known result from psychometric theory (see Lord and Novick, 1968, p. 70) that an observed correlation $r_{12}$ between two variates, which can only be measured with errors, is related to the correlation $\rho_{12}$ between their true scores by the expression

$$r_{12} = \rho_{12}(\rho_{11}\rho_{22})^{1/2}, \tag{7.18}$$

in which $\rho_{11}$ and $\rho_{22}$ are the reliability coefficients of the two tests respectively. This is Spearman's formula for 'correction for attenuation', mentioned in Chapter 1. (A reliability coefficient for a test is defined by

$$\rho_{11} = 1 - \sigma_e^2/\sigma_x^2,$$

in which $\sigma_x^2$ is the variance of the observed scores on the test and $\sigma_e^2$ the variance of the error scores.) As an example, suppose that for two tests the true correlation is $\rho_{12} = 0.5$ and that $\rho_{11} = 0.64$ and $\rho_{22} = 0.81$; then in sampling from the population the expected value of $r_{12}$, as given by (7.18), is only 0.36.

Well constructed tests, such as the standardized tests used by Emmett in his study, may have reliability coefficients as high as 0.95, but for *ad hoc* rating scales and other similar measuring instruments, reliability coefficients may fall well below this value.

When such variates are employed in a multiple regression analysis it is hardly surprizing that the outcome is disappointing. In recent years several attempts have been made (e.g. Cochran, 1970; Hodges and Moore, 1972; Lawley and Maxwell, 1973) to assess the effects of errors of measurement in such analyses, and some results given in the last of these three papers are now mentioned briefly, primarily because they do not require that prior estimates of error variances of the variates be available.

## Regression and factor analysis

Lawley and Maxwell, in their approach to the error problem, employ the factor model described in Chapter 5, and the equations for estimating factors given in the final section of Chapter 6. They express the regression equation in terms of the parameters of the factor model and show that certain useful inferences about the effects of measurement errors can then be drawn.

At the outset a factor analysis is performed using all $(1+p)$ variates and, for simplicity, we shall assume here that their correlation matrix is employed. Let the row vector $\boldsymbol{\lambda}_1'$ denote the loadings of the dependent variate, $y$, on the $k$ factors and let the $p \times k$ matrix $\boldsymbol{\Lambda}$ be the corresponding loading matrix for the $p$ independent variates. Denote their residual variances by the diagonal matrix $\boldsymbol{\Psi}$ of order $p$ and let the residual variance of $y$ be $\psi_1$. The sample correlation matrix, $\mathbf{S}$, for the independent variates is now replaced by the matrix $(\boldsymbol{\Lambda}\boldsymbol{\Lambda}' + \boldsymbol{\Psi})$, which contains the maximum likelihood estimates of the correlation coefficients in $\mathbf{S}$. Similarly the vector of sample correlations between $y$ and the $p$ independent variates is replaced by the row vector $\boldsymbol{\lambda}_1'\boldsymbol{\Lambda}'$. Since $y$ is in standardized form and the factors are taken to be unit normal variates the elements of $\boldsymbol{\lambda}_1'$ are the correlations between $y$ and the $k$ factors, denoted in Chapter 6 by the column vector $\mathbf{f}$ of order $k$. We may then write our prediction equation as

$$\hat{\mathbf{y}} = \boldsymbol{\lambda}_1'\hat{\mathbf{f}}_1, \tag{7.21}$$

in which $\hat{\mathbf{f}}$ is an estimate of $\mathbf{f}$.

Two prediction equations are available depending on whether we use (6.4) or (6.7) for estimating $\mathbf{f}$. When the former is used the estimates of $y$ are biased in the same sense that the estimates of the elements of $\mathbf{f}$ are biased, the bias being caused by the measurement errors in the independent variates. When (6.7) is employed the estimates of $y$ are unbiased despite the errors. The estimates of the multiple correlation coefficient in the two cases are given by

$$r_1^2 = \boldsymbol{\lambda}_1' \boldsymbol{\lambda}_1 - \boldsymbol{\lambda}_1'(\mathbf{I} + \boldsymbol{\Gamma})^{-1} \boldsymbol{\lambda}_1, \tag{7.22}$$

and

$$r_2^2 = (\boldsymbol{\lambda}_1' \boldsymbol{\lambda}_1)^2 / \{\boldsymbol{\lambda}_1'(\mathbf{I} + \boldsymbol{\Gamma}^{-1})\boldsymbol{\lambda}_1\} \tag{7.23}$$

respectively, in which $\boldsymbol{\Gamma} = \boldsymbol{\Lambda}'\boldsymbol{\Psi}^{-1}\boldsymbol{\Lambda}$, and it can be shown that $r_1^2 \geqslant r_2^2$. Hence unbiased estimates of $y$ are in general achieved at a price.

Lawley and Maxwell (1973) also show that if at least $k$ of the $p$ independent variates are measured without error the multiple correlation coefficient is not attenuated by errors in the remaining $(p-k)$ variates. Its square is then given by $\boldsymbol{\lambda}_1' \boldsymbol{\lambda}_1$, which is greater than either $r_1^2$ or $r_2^2$.

### Addendum

In the factor formulation of the regression equation it is assumed that the variates fulfil the conditions stated in Chapter 5 for minimizing the effects of specific factors, so that the estimates of the residual variances may be taken to be estimates of 'true' error variance. If in addition the sample size is sufficiently large for the sampling errors of these estimates to be ignored we may take $\boldsymbol{\Lambda}\boldsymbol{\Lambda}'$ to be the covariances of the independent variates after correction for error. It can then be shown that the regression equation for estimating $\mathbf{y}$ is

$$\hat{\mathbf{y}} = \boldsymbol{\lambda}_1' \boldsymbol{\Gamma}^{-1} \boldsymbol{\Lambda}' \boldsymbol{\Psi}^{-1} \mathbf{x}, \tag{7.24}$$

in which $\mathbf{x}$ is a column vector of order $p$ with elements $x_1$ to $x_p$. The multiple correlation coefficient with the $x$'s corrected for error, is given by $(\boldsymbol{\lambda}_1' \boldsymbol{\lambda}_1)^{1/2}$.

*Example* 7.2

In this example the independent variates consisted of scores for a sample of 220 boys on six school subjects obtained when they were of average age 16 years, the subjects being Gaelic, English, History, Arithmetic, Algebra and Geometry. The dependent variate consisted of scores on Algebra in an examination taken by the boys two years later, and the problem was to see how well these scores could be predicted from the scores on the original six subjects.

Table 7.1 Correlation coefficients for seven variates

| (1) | (2) | (3) | (4) | (5) | (6) | (7) |
|---|---|---|---|---|---|---|
| 1.000 | 0.249 | 0.284 | 0.146 | 0.495 | 0.572 | 0.539 |
| | 1.000 | 0.439 | 0.410 | 0.288 | 0.329 | 0.248 |
| | | 1.000 | 0.351 | 0.354 | 0.320 | 0.329 |
| | | | 1.000 | 0.164 | 0.190 | 0.181 |
| | | | | 1.000 | 0.595 | 0.470 |
| | | | | | 1.000 | 0.464 |
| | | | | | | 1.000 |

Table 7.2 Factor loadings and residual variances for 7 variates

| *Variate* | *Loadings* | | *Residual variances* |
|---|---|---|---|
| | I | II | |
| (1) | 0.704 | − 0.240 | 0.446 |
| (2) | 0.515 | 0.478 | 0.506 |
| (3) | 0.535 | 0.344 | 0.595 |
| (4) | 0.355 | 0.470 | 0.653 |
| (5) | 0.709 | − 0.145 | 0.476 |
| (6) | 0.754 | − 0.164 | 0.405 |
| (7) | 0.646 | − 0.134 | 0.565 |

The correlation matrix for all seven subjects, placing the dependent variate first and the independent variates in the order given above, are shown in Table 7.1.

A factor analysis of the correlation matrix for all 7 variates was carried out and after two factors had been fitted the test of 'goodness of fit' gave a value of chi-square of 10.34 which, with 8 degrees of freedom, is well below the 10% level of significance. The loadings on the two factors and the residual variances are shown in Table 7.2. This example is not entirely realistic, since it seems unlikely that the residual variances, which are somewhat large, represent true error variances. However, for purposes of illustration, we shall proceed as if they were.

The vector $\boldsymbol{\lambda}_1'$ is now $[0.704 \; -0.240]$; $\boldsymbol{\Lambda}$ contains the remaining factor loadings, while $\boldsymbol{\Psi}$ is the diagonal matrix having the last 6 residual variances as its diagonal elements. Using equation (7.21), with **f** estimated by (6.4), the regression weights are found to be

0.013 0.046 −0.023 0.250 0.317 0.193.

The first three are negligibly small, which is not surprizing, since one would not expect the scores on Gaelic, English and History to contribute much to the prediction of scores on Algebra. By use of (7.22) the multiple correlation coefficient is found to be 0.647.

When **f** is estimated using (6.7) the regression weights are found to be

−0.071 0.007 −0.097 0.356 0.453 0.273.

The first three are again very small, but the last three have increased considerably in magnitude. The latter six weights are those which give unbiased estimates of the dependent variate and since the multiple correlation coefficient (using 7.23) is found to be 0.604, which is only slightly less than the earlier estimate of 0.647, these weights appear preferable to the earlier set.

For this example $\boldsymbol{\lambda}_1' \boldsymbol{\lambda}_1 = 0.554$. Its square root is 0.744 and this is the estimate of what the multiple correlation coefficient would be if at least two of the independent variates were measured without error.

CHAPTER EIGHT

# Canonical Correlations

## Introduction

Canonical correlational analysis is a natural extension of multiple regression analysis. In the latter, as we have seen, the criterion to be predicted consisted of a single variate $y$, namely the dependent variate. But in some situations the criterion may be composite and consist of several $y$-variates. The problem then is to find a weighted sum of them which correlates maximally with a weighted sum of the predictor variates, the $x$'s of Chapter 7. The correlation coefficient derived is known as a canonical correlation coefficient, but for a given set of data, as we shall shortly see, several of these coefficients may exist, each independent of the other and they are then considered in decreasing order of magnitude.

It is not difficult to think of situations in which a canonical correlation analysis could be helpful for prediction purposes. For instance, in Example 7.1, it would have been interesting to have included as a criterion variate, in addition to children's academic success at secondary school, a measure of their participation in the corporate life of the school for it would also give an indication of the benefits they were deriving from being there. Indeed in personnel selection investigations in industry the criterion is generally composite in nature. For example, the success of a salesman may be measured by the value of the orders he acquires for his firm; but if he is working as a member of a sales team, overall success may depend on his ability to co-operate with his colleagues, realizing that one of them might succeed with a client with whom he had failed.

## The model

The classic papers on canonical correlational analysis are those by Hotelling (1935, 1936), but his work has been extended by others, in particular by Horst (1961), while Meredith (1964) has considered the effects of errors of measurement in the variates on parameter estimates. We shall consider the simplest case in which there are $p$ linearly independent but related variates. We shall assume that each pair of these is distributed approximately in a bivariate normal way and that, because of likely differences in metric, the scores on them are in standardized form. Let **S** be their correlation matrix, of order and rank $p$, and let it be partitioned as follows:—

$$\mathbf{S} = \begin{bmatrix} \mathbf{A} & \mathbf{C} \\ \mathbf{C}' & \mathbf{B} \end{bmatrix},$$

in which **A** is the matrix of correlations for $p_1$ predictor variates (the $x$'s), **B** is a matrix of correlations for $p_2$ criterion variates (the $y$'s), while **C** is the matrix of cross-correlations of order $p_1 \times p_2$ between the predictor and the criterion variates respectively $(p_1 + p_2 = p)$. At the outset the problem is to find a vector of weights, say $\mathbf{a}_1$, of order $p_1$ and a vector of weights, say $\mathbf{b}_1$, of order $p_2$ to apply to the predictor and criterion variates respectively such that the correlation coefficient between the composite scores thus obtained is a maximum. Once $\mathbf{a}_1$ and $\mathbf{b}_1$ have been determined, a further set of weights, say $\mathbf{a}_2$ and $\mathbf{b}_2$, independent of the first set may be sought and so on, but this additional aspect of the problem will shortly become clear.

Suppose that a vector of weights **b** were known, then the $p_2$ criterion variates could immediately be reduced to a composite variate $y^*$, given by

$$y^* = \Sigma b_i y_i, \qquad (i = 1 \ldots p_2) \tag{8.1}$$

whose covariances with the predictor variates are the elements of the column vector **Cb** and whose variance is **b′Bb**. The covariances may now be expressed as correlation coefficients by dividing each of the elements of **Cb** by the scalar quantity $(\mathbf{b}'\mathbf{B}\mathbf{b})^{1/2}$. The vector, **c**, of correlations between the predictor variates and the composite criterion variate, $y^*$, thus is

$$\mathbf{c} = \mathbf{C}\mathbf{b}/(\mathbf{b}'\mathbf{B}\mathbf{b})^{1/2}. \tag{8.2}$$

The partitioned matrix **S**, of order $p$, can now be replaced by the partitioned matrix **S***, of order $(p_1 + 1)$, where

$$\mathbf{S}^* = \begin{bmatrix} \mathbf{A} & \mathbf{c} \\ \mathbf{c}' & 1 \end{bmatrix}.$$

The data are now in a form suitable for the application of the multiple regression model of the last chapter and, by the use of (7.17), the multiple correlation coefficient between the $p_1$ predictor variates and $y^*$ is given by

$$r_{\mathrm{m}}^2 = \mathbf{c}'\mathbf{A}^{-1}\mathbf{c}, \tag{8.3}$$

and $r_{\mathrm{m}}$ is one of the canonical correlation coefficients required. On substituting the value of **c**, given by (8.2), in (8.3) and (for convenience) replacing $r_{\mathrm{m}}^2$ by $\lambda$, we get

$$\lambda = (\mathbf{b}'\mathbf{C}'\mathbf{A}^{-1}\mathbf{C}\mathbf{b})/(\mathbf{b}'\mathbf{B}\mathbf{b}). \tag{8.4}$$

To find the vector **b** which maximizes $\lambda$ we find $\partial\lambda/\partial\mathbf{b}$, (or $\partial\lambda/\partial\mathbf{b}'$) and set the result equal to zero.

In elementary calculus the derivative $\mathrm{d}y/\mathrm{d}x$ of a ratio of the form $y = v/u$, in which $v$ and $u$ are functions of a variable $x$, when set equal to zero, gives

$$v \cdot \mathrm{d}u/\mathrm{d}x = u \cdot \mathrm{d}v/\mathrm{d}x. \tag{8.5}$$

Hence if $v$ stands for the numerator in (8.4) and $u$ for the denominator, we obtain by (8.5) using $\partial\lambda/\partial\mathbf{b}'$ in the differentiation,

$$(\mathbf{b}'\mathbf{B}\mathbf{b})(2\mathbf{C}'\mathbf{A}^{-1}\mathbf{C}\mathbf{b}) = (\mathbf{b}'\mathbf{C}'\mathbf{A}^{-1}\mathbf{C}\mathbf{b})(2\mathbf{B}\mathbf{b}),$$

which, using (8.4) and dividing by 2, gives

$$\mathbf{C}'\mathbf{A}^{-1}\mathbf{C}\mathbf{b} = \lambda\mathbf{B}\mathbf{b},$$

The latter expression may be written in the form

$$(\mathbf{H} - \lambda\mathbf{B})\mathbf{b} = 0.$$

in which **H** stands for the matrix product $\mathbf{C}'\mathbf{A}^{-1}\mathbf{C}$, or, on premultiplication within the brackets by $\mathbf{B}^{-1}$, as

$$(\mathbf{B}^{-1}\mathbf{H} - \lambda\mathbf{I})\mathbf{b} = 0. \tag{8.6}$$

Hence $\lambda$ is a latent root of the matrix $\mathbf{B}^{-1}\mathbf{H}$. The rank of this matrix is $p_1$ or $p_2$, whichever is the smaller. Its largest latent root, $\lambda_1$, is the square of the largest canonical correlation coefficient (8.3) between the predictor and the criterion variates and the vector of weights $\mathbf{b}_1$ corresponding to it is the set of weights to be applied in (8.1). The companion set of weights, $\mathbf{a}_1$, for the predictor variates is then given by

$$\mathbf{a}_1 = \mathbf{A}^{-1}\mathbf{c}. \tag{8.7}$$

Further sets of weights and further canonical correlations are found by considering the remaining latent roots of $\mathbf{B}^{-1}\mathbf{H}$ in order of magnitude.

### A computational problem

The matrix $\mathbf{B}^{-1}\mathbf{H}$, whose latent roots and vectors have to be found, unfortunately, is not symmetric and since very powerful computer subroutines for finding the latent roots and vectors of symmetric matrices exist, it is convenient to replace $\mathbf{B}^{-1}\mathbf{H}$ by a matrix having the same latent roots but which is symmetric. To do this we return to equation (8.5) and express $\mathbf{B}$ in the form $\mathbf{B} = \mathbf{T}\,\mathbf{T}'$, in which $\mathbf{T}$ is a lower triangular matrix. The equation may now be written as

$$(\mathbf{H} - \lambda\,\mathbf{T}\,\mathbf{T}')\mathbf{b} = 0.$$

On pre-multiplying within the brackets by $\mathbf{T}^{-1}$ and post-multiplying by $\mathbf{T}'^{-1}$ we get

$$(\mathbf{T}^{-1}\mathbf{H}\,\mathbf{T}'^{-1} - \lambda\mathbf{I})\mathbf{b}^* = 0. \tag{8.8}$$

in which $\mathbf{T}^{-1}\mathbf{H}\,\mathbf{T}'^{-1}$ is a symmetric matrix which has the same latent roots as $\mathbf{B}^{-1}\mathbf{H}$. The transformation does however alter the elements of $\mathbf{b}$, but having found $b^*$ corresponding to one of the latent roots in (8.8) the required vector $\mathbf{b}$ of (8.6) is given by

$$\mathbf{b} = \mathbf{T}'^{-1}\mathbf{b}^*. \tag{8.9}$$

Verification of this expression is given in the last section of Chapter 3.

### Tests of significance

The matrix $\mathbf{B}^{-1}\mathbf{H}$ in (8.6) is of order and rank $p^*$, where $p^*$ is the smaller of $p_1$ and $p_2$, hence it has $p^*$ latent roots. Assuming that they are distinct then the analysis provides $p^*$ independent canonical correlation coefficients; but some or all of them may be too small to be of any practical value. As an aid to their assessment Bartlett (1947) has supplied an approximate criterion for testing their significance. It is

$$X^2 = -\{(N-1) - \tfrac{1}{2}(p+1)\}\log_e(1-\lambda_i), \tag{8.10}$$

in which $N$ is the sample size, $p$ is the total number of variates and $\lambda_i$ is the $i$th latent root of $\mathbf{B}^{-1}\mathbf{H}$. For moderately large $N$ the criterion is distributed approximately as chi-square, with $(p+1-2i)$ degrees of freedom. But, in using it, it is well to remember that the size of a correlation coefficient is of more practical importance than its significance level, however, convincing the latter may prove to be.

*Example* 8.1

A sample of 130 women students in a medical school took the subjects (1) Chemistry, (2) Zoology and (3) Botany in their first medical examination, and (4) Anatomy and (5) Physiology in their second medical examination a year later. The examiners wished to determine the overall relationship between the students' scores in the two examinations. For each examination paper the maximum score was 100 and the means and standard deviations of the scores obtained by the students are given in Table 8.1.

The means and standard deviations vary considerably from subject to subject and this variation is no doubt due in part to differences in the relative difficulties of the examination papers and to differences in the scoring methods of the examiners. To help in overcoming these extraneous factors the correlations rather than the covariances of the variates were employed in the subsequent analysis of the data, and the correlation matrix is given in Table 8.2.

The correlation coefficients are all positive as one would expect.

Table 8.1 Means and standard deviations of students' scores

| | *Chemistry* | *Zoology* | *Botany* | *Anatomy* | *Physiology* |
|---|---|---|---|---|---|
| Mean | 49.41 | 46.71 | 58.75 | 50.87 | 51.15 |
| S.D. | 14.36 | 9.86 | 12.05 | 11.58 | 9.49 |

Table 8.2 The correlation matrix for the 5 subjects

| (1) | (2) | (3) | (4) | (5) |
|---|---|---|---|---|
| 1.000 | 0.652 | 0.660 | 0.661 | 0.566 |
| 0.652 | 1.000 | 0.593 | 0.509 | 0.499 |
| 0.660 | 0.593 | 1.000 | 0.579 | 0.573 |
| 0.661 | 0.509 | 0.579 | 1.000 | 0.714 |
| 0.566 | 0.499 | 0.573 | 0.714 | 1.000 |

They are also remarkably consistent in magnitude and since the highest observed correlation between a predictor and a criterion variate is 0.661 (between Chemistry and Anatomy) we know immediately that the largest canonical correlation will be, at least, of this magnitude.

The matrix in Table 8.2 is partitioned to correspond to the symbolic notation used earlier in the chapter, namely,

$$\begin{array}{c|c} \mathbf{A} & \mathbf{C} \\ \hline \mathbf{C'} & \mathbf{B} \end{array}.$$

At the outset we calculate $\mathbf{A}^{-1}$, $\mathbf{H} = \mathbf{C'}\mathbf{A}^{-1}\mathbf{C}$ and the lower triangular matrix $\mathbf{T}$, where $\mathbf{T}\,\mathbf{T'} = \mathbf{B}$, together with its inverse $\mathbf{T}^{-1}$. They are:—

$$\mathbf{A}^{-1} = \begin{bmatrix} 2.174 & -0.872 & -0.917 \\ -0.873 & 1.894 & -0.548 \\ -0.917 & -0.548 & 1.931 \end{bmatrix}, \mathbf{H} = \begin{bmatrix} 0.476 & 0.430 \\ 0.430 & 0.401 \end{bmatrix},$$

$$\mathbf{T} = \begin{bmatrix} 1.000 & \\ 0.714 & 0.700 \end{bmatrix}, \quad \mathbf{T}^{-1} = \begin{bmatrix} 1.000 & \\ -1.020 & 1.429 \end{bmatrix}.$$

We then find, for use in equation (8.8), the matrix

$$\mathbf{T}^{-1}\mathbf{H}\,\mathbf{T'}^{-1} = \begin{bmatrix} 0.476 & 0.129 \\ 0.129 & 0.060 \end{bmatrix}.$$

Its latent roots are $\lambda_1 = 0.5128$ and $\lambda_2 = 0.0232$, with corresponding latent column vectors

$$\mathbf{b}_1^* = \begin{bmatrix} 1.0000 \\ 0.2853 \end{bmatrix} \quad \text{and } \mathbf{b}_2^* = \begin{bmatrix} -0.2853 \\ 1.0000 \end{bmatrix},$$

and, using (8.9), we get the required vectors, namely

$$\mathbf{b}_1 = \begin{bmatrix} 0.7090 \\ 0.4077 \end{bmatrix} \quad \text{and } \mathbf{b}_2 = \begin{bmatrix} -1.3053 \\ 1.4290 \end{bmatrix},$$

Of the two latent roots, the first is dominant and yields a canonical correlation coefficient of $\sqrt{0.5128} = 0.716$. Its significance is hardly in doubt, but may be tested by (8.10), which gives the value

$$X^2 = -126 \log_e(1 - 0.5128) = 90.61,$$

based on 4 degrees of freedom. For the second latent root $X^2 = 2.96$ which, with 2 degrees of freedom, does not reach an acceptable level of significance and consequently this root may be neglected.

To find the vector of weights $\mathbf{a}_1$ for the predictor variates, corresponding to $\mathbf{b}_1$, we substitute the latter for $\mathbf{b}$ in (8.2) to find $\mathbf{c}_1$, and then use (8.7). The result is the column vector

$$\mathbf{a}_1 = \{0.420 \quad 0.103 \quad 0.282\}.$$

In assessing the results it is helpful to scale the weights so that the largest element in $\mathbf{a}_1$ and $\mathbf{b}_1$ separately is unity. They then become:—

| *Chemistry* | *Zoology* | *Botany* | : | *Anatomy* | *Physiology* |
|---|---|---|---|---|---|
| 1.000 | 0.245 | 0.671 | : | 1.000 | 0.575 . |

In the weighted sums of standardized scores on the two sets of variates, Chemistry and Anatomy respectively are seen to play the leading rôles. But it is not without significance that Zoology and Physiology, which play the lesser rôles, are the two variates whose standard deviations, Table 8.1, are lowest, and one is left with the impression that had the examination papers in these two subjects discriminated better between the students the relative rôles of the variates might well have been different. If all variates are given equal weights the correlation coefficient between the straight sums of standardized scores on the two sets of variates is found to be 0.701, which is only slightly less than the value 0.716 found for the canonical coefficient. Hence, despite the vagaries of the examination process, the overall correlation between the students' performance in the two examinations is seen to be about 0.7. There is little doubt that this value could be increased by an improvement in examining techniques to give more reliable scores.

## Inter-rater agreement

Canonical correlation analysis can also be usefully employed in assessing the degree of agreement between two judges who independently rate a sample of individuals with respect to a set of characteristics. For example, two psychiatrists $P_1$ and $P_2$ interviewed a sample of patients and rated them on a 5-point scale on each of the three variates:

$$x_1 = \text{lack of concentration},$$
$$x_2 = \text{despondency},$$
$$x_3 = \text{anxiety}.$$

In this instance the symmetric matrices **A** and **B** consist of the intercorrelations of the three variates for the two psychiatrists respectively, while the matrix **C** contains the correlations between the scores given by one psychiatrist and those given by the other. The full results are not reported here but the matrix **C**, which was found to be

| | | $P_2$ | | |
|---|---|---|---|---|
| | | $x_1$ | $x_2$ | $x_3$ |
| $P_1$ | $x_1$ | 0.851 | 0.589 | 0.521 |
| | $x_2$ | 0.402 | 0.916 | 0.692 |
| | $x_3$ | 0.198 | 0.644 | 0.812 |

is worth special comment. The entries in its main diagonal are the correlations between the two psychiatrists' scores on the three variates. The highest correlation is 0.916 for $x_2$ and we know immediately that the largest canonical correlation coefficient cannot be less than this. Moreover, since the corresponding correlations for $x_1$ and $x_3$ are also relatively large, overall agreement between the psychiatrists is remarkably good.

On proceeding with the analysis the first canonical correlation coefficient was found to be 0.957, the corresponding weights for the variates being

| $P_1$ | | | $P_2$ | | |
|---|---|---|---|---|---|
| $x_1$ | $x_2$ | $x_3$ | $x_1$ | $x_2$ | $x_3$ |
| 0.852 | 1.000 | 0.266 | 0.630 | 1.000 | 0.457 |

Inspection of the results shows that in each case $x_2$ (despondancy) gets the highest weight and $x_1$ (lack of concentration) the second highest.

Two additional points are worth mentioning. The first latent root $\lambda_1$ of the matrix $\mathbf{T}^{-1}\mathbf{H}\mathbf{T}^{-1}$ for these data accounted for only 49% of its trace, consequently at least one further root of the matrix should be examined. But additionally we should note that our correlational analysis of the data, however satisfactory the outcome, refers only to concomitant agreement between the scores alotted by the psychiatrists to the patients. It tells us nothing about possible bias in the results in the sense that the means of the variates may differ from one psychiatrist to the other and they should be examined separately.

## General comments

Canonical correlational analysis has been described (Bartlett, 1948) as 'external factor analysis', though a more exact description would be 'external component analysis', since the problem of the *communalities* of variates is not involved. It is 'external' in the sense that the hypothetical variates derived from each of the two distinct sets of observed variates are not *principal* components of these sets. Rather they are components which, in pairs, are related to each other in a regression sense (see 8.7) from one set of variates to the other and indeed seldom correspond to principal components themselves or correlate highly with them.

Initially, the hope was entertained, that canonical correlational analysis would prove useful in revealing relationships between sets of variates of different types such as cognitive and physical, or cognitive and temperamental variates; but as Bartlett notes (and through no fault of the technique) these hopes were not fulfilled as the correlations between variates of different types are generally small. But the technique is still likely to prove useful in prediction studies in which the criterion is multivariate in form: we recall that Hotelling's original paper on the subject has the title 'The most predictable criterion'.

CHAPTER NINE

# Discriminant Function and Canonical Variate Analysis

## Introduction

Suppose that for some purpose the members of a population can usefully be divided into fairly distinct groups with respect to certain characteristics which, to a greater or lesser extent, they possess. For example, it is common practice amongst psychiatrists to classify their patients into categories, such as schizophrenics, affective psychotics, neurotics and so on, on the basis of symptoms observed in routine clinical examinations, for this is helpful in prescribing appropriate treatment. Now, given a sample of individuals allocated to groups in such a manner, one might wish to check on the validity of the clinician's allocation by the use of laboratory tests and other objective measures of characteristics known to be involved, and on the basis of such measures to derive decision rules for allocating new individuals to one or other of the groups so that the number of misclassifications is minimized. This is one type of situation in which discriminant function and canonical variate analysis can be helpful, for the ensuing discussion between clinician and statistician about any discrepancies in their respective allocation of patients may throw light on the best sets of variables to use and the most appropriate symptoms to note to improve classification.

In other situations a time factor may be involved and the problem becomes one of predicting future group membership of an individual based on measurements made on him at a time before the special group characteristics themselves can be observed. The problem discussed in Example 6.1 provides an illustration. Once children have reached the age when they should normally be able to read, their reading ability can be tested and they can broadly be classified into groups, as in that example, with respect to it. But, with remedial teaching in mind, it would be useful to be able to identify as an earlier age, using tests of general cognitive ability and attainment, children who were likely to experience difficulty in learning to read. This problem, and that concerned with the classification of psychiatric patients, is discussed later in the present chapter after the statistical models required for discrimination have been presented.

## The case of two groups

Denote the characteristics to be measured by the $p$ variates $X_1$ to $X_p$. We wish to find a weighted function of these variates which best discriminates between the two groups. One approach to the problem is to use the multiple regression model of Chapter 7. The dependent variate $Y$ is now treated as a dummy variate which takes the value (1) if an individual belongs to one group and (0) if he belongs to the other. The $X$'s are then used as predictor variates and a linear function of them is found which predicts the $Y$ values as well as possible.

If there are $n_1$ individuals in the first group and $n_2$ in the second the model formally resembles that of equation (7.2) with $n = n_1 + n_2$ so that the two groups are combined. The vector of weights is given by (7.5) and the function we require takes the form (7.7), though the term $a$ in it may be dropped as it is the same for both groups. Thus we obtain a function of the form

$$Y = b_1X_1 + b_2X_2 + \ldots + b_pX_p. \qquad (9.1)$$

If in this function we replace the $X$-values by their means in the first group we get a weighted mean score, say $Y_1$, for it. In a similar manner we find a weighted mean score, say $Y_2$, for the second group. Under the assumptions that the $X$'s have a multivariate normal distribution in each group and that corresponding variances and covariances are the same in the two groups, it can be shown that the $Y$-value given by

$$Y^* = (Y_1 + Y_2)/2, \qquad (9.2)$$

when used as the point of separation between the groups, minimizes the misclassification of individuals.

In practice the model has certain obvious limitations for it assumes that the $n_1$ and $n_2$ individuals in the respective groups were correctly classified in the first instance, and that the prior probabilities of group membership are equal. Since prior probabilities are seldom known the results given by using $Y^*$, as a criterion value, can only be assessed in terms of the overall percentage of misclassifications it yields and in terms of the relative seriousness attached by the user to the two types of misclassification involved. It is also clear that the percentage of individuals misclassified will in general be underestimated if it is based on the sample used in finding the discriminant function, since the function is optimal for that particular set of data and may not have good general validity.

The original papers on discriminant functions were those by Fisher (1936) and Welch (1939) and were concerned with the two-group problem, but methods for dealing with more than two groups simultaneously were later given by Rao (1948). These more general methods usually involve more than one function of the $X$'s and these functions are commonly called *canonical variates.* Discriminant functions are now widely used in research work. Armitage (1971, Chapter 10) gives a good review of the simpler techniques available and interesting applications are reported by Radhakrishna (1964), Healy (1965), and many others. The techniques have also been found useful in situations in which the variates employed are dichotomous in form (Claringbold, 1958; Maxwell, 1961). We shall now consider in detail a model which is applicable in the case of two or more groups.

## Canonical variate analysis

Suppose that we have $g$ groups with $n_i$ individuals in the $i$th group. Let $X_j (j = 1, \ldots, p)$ be the variates being used to discriminate between them and assume that in each group the variates have a multivariate normal distribution with the same variance-covariance matrix but with differing means. Let the symmetric matrix $\mathbf{T}$, of order $p$, represent the corrected sums of squares and cross-products of the variate scores for the total sample of $n = \Sigma n_i$ individuals, $(i = 1, \ldots, g)$. Let $\mathbf{W}_i$ be a similar matrix for the $i$th group and let

$$\mathbf{W} = \mathbf{W}_1 + \mathbf{W}_2 + \ldots + \mathbf{W}_g. \tag{9.2}$$

$\mathbf{W}$ is then the pooled 'within groups' sums of squares and cross-products, while the 'between groups' sums of squares and cross products is

$$\mathbf{B} = \mathbf{T} - \mathbf{W}. \tag{9.3}$$

Assume a linear function of the $X$'s, as in (9.1), which combines them into a composite variate $Y$. On this variate the sums of squares 'between' and 'within' groups are $\mathbf{b}'\mathbf{B}\mathbf{b}$ and $\mathbf{b}'\mathbf{W}\mathbf{b}$ respectively, where $\mathbf{b}$ is a column vector of weights. The ratio

$$\lambda = (\mathbf{b}'\mathbf{B}\mathbf{b})/(\mathbf{b}'\mathbf{W}\mathbf{b}), \tag{9.4}$$

in the ratio of 'between' to 'within' sums of squares for the $g$ groups on the composite variate $Y$. If $\lambda$ is multiplied by $(n-g)/(g-1)$ we get the $F$-ratio for a one-way analysis of variance of the $Y$-scores for the $g$ groups based on $(g-1)$ and $(n-g)$ degrees of freedom. Hence if we find the vector $\mathbf{b}$ which maximizes $\lambda$ it will provide us with a set of weights to substitute in (9.1) to maximize mean group differences.

To maximize $\lambda$ we find $\partial\lambda/\partial\mathbf{b}'$ and set the result equal to zero. We follow closely the procedure described in Chapter 8 and, using (8.5) with $v = \mathbf{b}'\mathbf{W}\mathbf{b}$ and $u = \mathbf{b}'\mathbf{B}\mathbf{b}$, we get

$$(\mathbf{b}'\mathbf{W}\mathbf{b})(2\mathbf{B}\mathbf{b}) = (\mathbf{b}'\mathbf{B}\mathbf{b})(2\mathbf{W}\mathbf{b}),$$

which, on dividing by $2\mathbf{b}'\mathbf{W}\mathbf{b}$, is

$$\mathbf{B}\mathbf{b} = \lambda\mathbf{W}\mathbf{b},$$

and gives

$$(\mathbf{B} - \lambda\mathbf{W})\mathbf{b} = 0 \tag{9.5}$$

Hence $\lambda$ is a latent root of the matrix $\mathbf{W}^{-1}\mathbf{B}$ and $\mathbf{b}$ the corresponding latent column vector. Since $\mathbf{W}^{-1}\mathbf{B}$ is not symmetric we again follow the procedure described in the last section of Chapter 3 and again in Chapter 8 and replace it by a symmatric matrix having the same latent roots.

Given that the variates are linearly independent and $(n-g) \geqslant p$ the matrix $\mathbf{W}$ is of full rank $p$ and its inverse exists. The matrix $\mathbf{B}$, of order $p$, has a rank of $p$ or $(g-1)$, whichever is the smaller, since there are only $(g-1)$ independent comparisons between group means; hence the rank of $\mathbf{W}^{-1}\mathbf{B}$ is $p^*$ where $p^*$ is the smaller of $p$ and $(g-1)$.

## Interpretation

Provided that the $p^*$ latent roots of the matrix $\mathbf{W}^{-1}\mathbf{B}$ are distinct the corresponding latent vectors will provide $p^*$ independent canonical variates of the form (9.1). This implies that the means of the groups lie in a $p^*$-dimensional space. But variation in some of the dimensions may be too small to merit attention, and a criterion for testing the significance of a latent root $\lambda_j (j = 1, \ldots, p^*)$ is

$$X^2 = \{n - 1 - \tfrac{1}{2}(p + g)\} \log_e(1 + \lambda_j). \tag{9.6}$$

For moderately large $n$ it is distributed as chi-square with $(p + g - 2j)$ degrees of freedom.

In interpreting the results of an analysis the best procedure is to calculate a score for each individual on each canonical variate. Group means on each variate may then be compared using tests of significance if desired. More importantly the scores may be plotted in diagrams and the extent of overlap between groups noted. If only one of the latent roots is significant, or if only two groups are involved in the analysis, then the results lie on a single dimension and their interpretation should be straightforward. When two dimensions are required the results can be displayed in the form of a scattergram and overlap between groups noted. In the case of three or more dimensions a visual appraisal of the results is more difficult, but it is often helpful to plot the results taking the dimensions in pairs. Some computer programs calculate the likelihood, under multivariate normal assumptions, of each individual belonging to each group: the individual can then be allocated to the group for which his likelihood is greatest, and misclassification noted. This procedure, however, is sensitive to departures from normality in regions of overlap, and should be used with care.

## Scaling of scores on the observed variates

From an arithmetic viewpoint it is of interest to note that the scores on the canonical variates, relative to each other, are unchanged by linear changes of scale in the observed variates. Consequently, if the variates have different matrices, possibly with scores of very different magnitudes on each, it may be convenient at the outset to multiply by arbitrary constants to make the variances of the variates roughly equal.

Let $\mathbf{D}$ be a diagonal matrix whose diagonal elements are the arbitrary scaling constants decided upon. The ratio given by (9.4) now becomes

$$\lambda = (\mathbf{u}'\mathbf{DBDu})/(\mathbf{u}'\mathbf{DWDu}), \tag{9.7}$$

in which $\mathbf{u}$ replaces $\mathbf{b}$ and is the column of weights, of order $p$, to be determined to maximize $\lambda$. On repeating the algebra of the earlier section we find that (9.5) is replaced by

$$(\mathbf{B} - \lambda\mathbf{W})\mathbf{D}\mathbf{u} = 0. \tag{9.8}$$

Hence the original vector of weights, $\mathbf{b}$, is replaced by the vector $\mathbf{Du}$, that is

$$\mathbf{u} = \mathbf{D}^{-1}\mathbf{b}. \tag{9.9}$$

It follows that the scaling of the observed variates by $\mathbf{D}$ has the effect of scaling the weights by $\mathbf{D}^{-1}$ and the scores on the canonical variates remain unchanged. From this result it is clear that the interpretation of a canonical variate in terms of the relative magnitudes of the weights of the variates is justified only in cases in which all the variates have the same metric, for in other cases the weights depend on the scale units and these may be quite arbitrary.

## Scaling the canonical variates

The previous section was concerned with the relative weights of the observed variates in a single canonical variate. But the relative variances of the scores on different canonical variates are also of interest. In general we might want each variate to reflect the relative magnitude of the latent root of $\mathbf{W}^{-1}\mathbf{B}$ to which it referred. This can easily be achieved by scaling the elements of the vector of weights. The pooled 'within groups' covariance matrix is estimated by

$$\mathbf{S} = \mathbf{W}/(n - g), \tag{9.10}$$

and the variance of the $j$th canonical variate by

$$v_j = \mathbf{b}_j'\mathbf{S}\mathbf{b}_j, \tag{9.11}$$

where $\mathbf{b}_j$ is the $j$th vector of weights corresponding to the $j$th latent root $\lambda_j$. We want $v_j = \lambda_j$, hence the elements of $\mathbf{b}_j$ must be multiplied by $(\lambda_j/v_j)^{1/2}$.

Another common procedure is to scale the vector $\mathbf{b}_j$ so that $v_j = c$, where $c$ is an arbitrary constant and is the same for all $j$. The variances of the canonical variates are now all equal and if the scores on any two of them are plotted in a diagram, and the positions of the mean points for the groups noted, confidence regions for these points can be represented by circles. For the $i$th group the radius of the circle is given by the square root of

$$\{2cF(n_i - 1)\}/\{n_i(n_i - 2)\},$$

in which the value of $F$, based on 2 and $(n_i - 2)$ degrees of freedom, is taken from the $F$-distribution. For example, for the 95% confidence region we would use the 5% tabulated value of $F$. (For a discussion of confidence regions of mean vectors see Morrison, 1967, p. 121).

*Example* 9.1

A psychiatrist interested in a possible sub-classification of schizophrenic patients interviewed a random sample of 67 and, on the basis of their case histories and his observation of symptoms which they displayed in interview, allocated them to one or other of four categories, I to IV, the numbers in each being 22, 19, 13 and 13 respectively. He then asked a statistician to check the allocation by a quantitative analysis, and supplied him with ratings for the patients on each of 9 scales labelled respectively:

(1) Depression,
(2) Somatic symptoms,
(3) Auditory hallucinations,
(4) Delusions of reference,
(5) Delusions of grandeur,
(6) Thought disorder,
(7) Abnormalities of motor behaviour,
(8) Blunting affect,
(9) Speech disorder.

Since the psychiatrist undoubtedly had these 9 variables, amongst others, in mind when he made his allocation of the patients to groups, the classic conditions for the use of discriminant functions are not truly met, yet the problem posed is of interest in its own right and may be tackled by a canonical variate analysis.

The data were analysed by computer. The matrix $\mathbf{W}^{-1}\mathbf{B}$ of order 9 was calculated. It has rank three, since $g - 1 = 3$, and its three latent roots (given below) were all found to be highly significant:

| | *Root* | *d.f.* | $X^2$ |
|---|---|---|---|
| $\lambda_1$ | 3.979 | 11 | 95.51 |
| $\lambda_2$ | 1.743 | 9 | 60.04 |
| $\lambda_3$ | 1.187 | 7 | 46.57 |

Using the latent vectors corresponding to these roots three canonical variate functions $y_1$ $y_2$ and $y_3$, were set up and a score for

Table 9.1 Mean scores on 3 canonical variates

| *Categories* | *Variate means* | | |
|---|---|---|---|
| | $y_1$ | $y_2$ | $y_3$ |
| I | −4.27 | 1.50 | 0.22 |
| II | 1.06 | −1.75 | 1.35 |
| III | 6.59 | 1.80 | −0.36 |
| IV | −0.93 | −1.79 | −1.99 |

each patient on each variate calculated. The mean scores for the four categories are shown in Table 9.1.

The means lie in a three-dimensional space and inspection suggests that variate $y_1$ may be useful in differentiating between categories I and III, variate $y_3$ in differentiating between categories II and IV, while $y_2$ contrasts categories I and III with II and IV. Further insight into the results is given if the canonical variates are taken in pairs, as orthogonal axes, and the mean scores for the categories plotted. The plot, using $y_1$ and $y_2$, is shown in Figure 9.1. It shows the wide contrast between categories I and III, and since their respective means scores on $y_3$ are small these categories may be

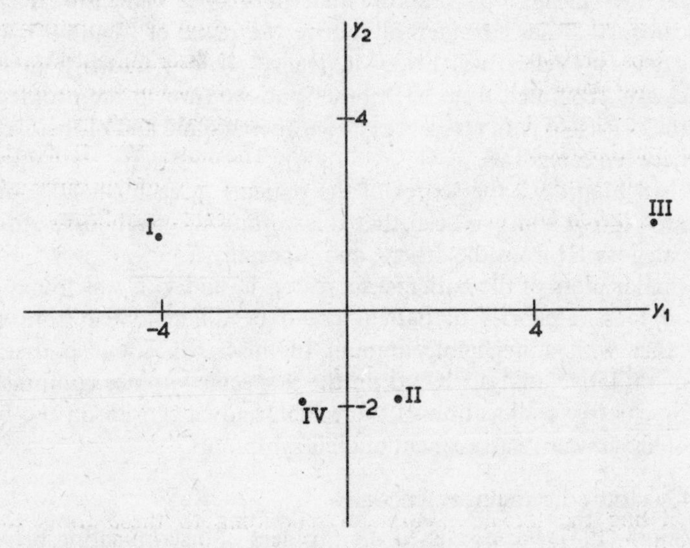


Fig. 9.1. *Means for Categories on canonical variates* $y_1$ *and* $y_2$.

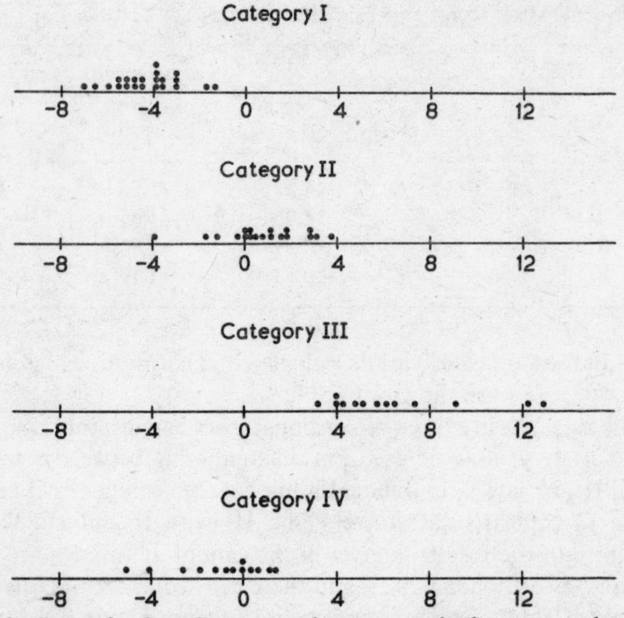


Fig. 9.2. *Scores for the 4 Categories of patients on the first canonical variate.*

taken to lie virtually in this plane. In the figure, the mean points for categories II and IV are not well differentiated but inspection of their means on $y_3$ shows that in a three dimensional space category II lies considerably above the plane of Figure 9.1 while category IV lies well below it. Hence all four means points are clearly separated from each other and we turn to the problem of the scatter of patients' scores about these points and of the overlap between categories.

In Figure 9.2 the scores of the patients in each category on $y_1$ are plotted and it is seen that this variate effectively differentiates category III from the others, and especially from category I. From similar plots of the patients' scores on $y_2$ and $y_3$ it was found that all four categories of patient could be differentiated from each other with a negligible amount of misclassification so that the quantitative analysis based on the 9 specific variates confirms the psychiatrist's allocation of the patients to categories on the basis of his subjective assessment of their symptoms.

**Quadratic discriminant functions**

Smith (1947) has discussed the problem of discriminating between two populations when their covariance matrices for the variates to be employed are not equal. Given multivariate normality in each population he has shown that the best discriminant function is then quadratic rather than linear in form. His method, which can readily be extended to cases involving more than two populations (e.g. Lubin, 1950, p. 98), has not been widely used in the past no doubt because it involves heavy calculations. But now that computing is no longer an obstacle a good case can be made for the more general use of quadratic functions, since the assumption of equal covariance matrices otherwise required is often not justified.

The quadratic expression given by Smith (1947, bottom of p. 275) can neatly by expressed in terms of two quadratic forms and some constants. It is derived by taking the natural logarithm of the ratio of the multivariate density function (for a definition see Morrison, 1967, p. 81) of one population ($P_1$) to that of the other ($P_2$), and is given below in terms of sample estimates of the population mean vectors and covariance matrices, since their true values are seldom available.

For $p$ variates, denote the estimates of the population means by the column vectors $\bar{\mathbf{x}}_1$ and $\bar{\mathbf{x}}_2$ and of the covariance matrices by $\mathbf{S}_1$ and $\mathbf{S}_2$ respectively. Let the random vector $\mathbf{x}$ stand for the scores of an individual drawn from the composite population (composed of $P_1$ and $P_2$) and construct the vectors

$$\mathbf{d}_1 = (\mathbf{x} - \bar{\mathbf{x}}_1) \text{ and } \mathbf{d}_2 = (\mathbf{x} - \bar{\mathbf{x}}_2).$$

The quadratic function for discriminating between the two populations is then given by

$$L = -\{\mathbf{d}_1' \mathbf{S}_1^{-1} \mathbf{d}_1 + \log_e |\mathbf{S}_1|\} + \{\mathbf{d}_2' \mathbf{S}_2^{-1} \mathbf{d}_2 + \log_e |\mathbf{S}_2|\} \quad (9.12)$$

To use the function, given $\bar{\mathbf{x}}_1, \bar{\mathbf{x}}_2, \mathbf{S}_1$ and $\mathbf{S}_2$, we find the vectors $\mathbf{d}_1$ and $\mathbf{d}_2$ for an individual with observed scores $\mathbf{x}$, calculate his $L$-score using (9.12), and assign him to $P_1$ if this score is negative and to $P_2$ if it is (zero or) positive. When relatively small samples are involved a shortcut is to calculate

$$L^* = -\mathbf{d}_1' \mathbf{S}_1^{-1} \mathbf{d}_1 + \mathbf{d}_2' \mathbf{S}_2^{-1} \mathbf{d}_2 \quad (9.13)$$

and to find the 'best' criterion point (say $L_0^*$) by inspection of the distributions of the $L^*$-scores for the two samples. $L^*$ differs from $L$ only by a constant.

When expression (9.12) is used to assign an individual, drawn randomly from the composite population, to $P_1$ or $P_2$ it is assumed that, prior to observing **x**, he is equally likely to belong to either. If this is not the case and prior probabilities of membership of $P_1$ and $P_2$ are known these can be used to improve allocation (see Welch, 1939).

*Example* 9.2

The greater efficiency of a quadratic over a linear discriminant function, when the covariance matrices are not equal, is well illustrated by the data of Example 6.1. Using a linear function to discriminate between 'good' and 'poor' readers and a criterion value $Y^*$, as defined in (9.2), the classification of children shown in Table 9.2 was obtained.

Table 9.2 Classification of children given by a linear discriminant function

| | *Below* $Y^*$ | *Above* $Y^*$ | *Total* |
|---|---|---|---|
| Good readers—$P_1$ | 16 | 59 | 75 |
| Poor readers—$P_2$ | 51 | 24 | 75 |

From the table it is seen that 16 of the 75 good readers are misclassified and 24 of the 75 poor readers, a total of 40 misclassifications. When the quadratic function (9.13) was employed, the classification of the children was that shown in Table 9.3, and it is seen that the total number of misclassifications is reduced to $9 + 13 = 22$.

In conclusion, a word of help is needed for any clinical psychologist who might wish to use these results, and who does not have ready access to a computer. The ten variates employed are the subtests of the well-known Wechsler Pre-School and Primary Scale of Intelligence. Comparison of the $L$-scores for 'poor' readers,

Table 9.3 Classification of children given by a quadratic discriminant function

| | *Above* $L_0$ | *Below* $L_0$ | *Total* |
|---|---|---|---|
| Good readers | 9 | 66 | 75 |
| Poor readers | 62 | 13 | 75 |

for the sample of children used in this study, with their average scores on all 10 of the Wechsler subtests indicated that a child whose average score was approximately 9.0 or less was likely to be at risk where learning to read was concerned. But this criterion is not to be relied upon implicitly, as the correlation between the *L*-scores and the average Wechsler scores is not high (about 0.7).

**Comparing covariance matrices**

Let $\mathbf{S}_1$ and $\mathbf{S}_2$ be the usual sample covariance matrices, based on $n_1$ and $n_2$ degrees of freedom respectively, for $p$ variates in two populations for which the true matrices are $\boldsymbol{\Sigma}_1$ and $\boldsymbol{\Sigma}_2$. Then a criterion for testing the hypothesis that $\boldsymbol{\Sigma}_1 = \boldsymbol{\Sigma}_2$ is

$$n \log_e |\mathbf{S}| - n_1 \log_e |\mathbf{S}_1| - n_2 \log_e |\mathbf{S}_2|, \qquad (9.14)$$

where $n = n_1 + n_2$ and where

$$\mathbf{S} = (n_1 \mathbf{S}_1 + n_2 \mathbf{S}_2)/(n_1 + n_2).$$

Under the hypothesis the criterion is distributed approximately as chi-squared with $\frac{1}{2}p(p+1)$ degrees of freedom. For a discussion of the test see Box (1949).

CHAPTER TEN

# The Analysis of Contingency Tables

## Introduction

The chi-square test is widely used for detecting association in a contingency table, but when the table is large, and a significant association is indicated by the test, difficulty is frequently experienced in interpreting the result. Well-known approaches to the problem (see Maxwell, 1964, Chapters 3 and 4) are to partition the overall chi-square value into additive components, when prior justification for doing so exists and the partitioning is likely to assist interpretation. A complimentary approach to the problem, first suggested by E. J. Williams (1952), partitions the overall chi-square value into additive components in terms of the latent roots of a matrix derived from the contingency table. As this approach is helpful in practice it is described briefly below, and illustrative examples given.

## Notation

Consider a sample of $n$ individuals classified by two attributes so that we have a contingency table with $p$ rows and $q$ columns and, for convenience, assume that $p \geqslant q$. Denote the contingency table by the matrix $\mathbf{A} = [a_{ij}], i = 1, 2, \ldots, p$ and $j = 1, 2, \ldots, q$, and let $a_{i\cdot}$ and $a_{\cdot j}$ be the total of its $i$th row and $j$th column respectively. The usual overall criterion for testing for association between the attributes is

$$X^2 = n\left[\sum_i \sum_j \frac{a_{ij}^2}{a_{i\cdot}a_{\cdot j}} - 1\right] \tag{10.1}$$

which is referred to the chi-square distribution with $(p-1)(q-1)$ degrees of freedom. For examples, see Maxwell, 1964, Chapter 2.

It is easy to verify that the first term within the brackets in (10.1) is the trace of the $q \times q$ symmetric matrix

$$\mathbf{T} = \mathbf{G}'\mathbf{G}, \tag{10.2}$$

in which $\mathbf{G} = [g_{ij}]$ is a matrix of the same order as $\mathbf{A}$ and formed from it by setting

$$g_{ij} = a_{ij}/(a_{i.}a_{.j})^{1/2},$$

$\mathbf{G}'$ being the transpose of $\mathbf{G}$. Having calculated $\mathbf{T}$, the criterion (10.1) may be written as

$$X^2 = n(\text{tr.}\,\mathbf{T} - 1), \tag{10.3}$$

in which 'tr.$\mathbf{T}$' stands for the trace of $\mathbf{T}$.

In the special case in which there is no association between the attributes, and sampling errors are absent, $X^2$ has the value zero so that

$$\text{tr.}\mathbf{T} = 1. \tag{10.4}$$

This is the case in which each element $a_{ij}$ of $\mathbf{A}$ is equal to its expected value $(a_{i.}a_{.j})/n$, and it can then be verified that

$$\mathbf{T} = \mathbf{a}\mathbf{a}', \tag{10.5}$$

in which $\mathbf{a}$ is a column vector of order $q$ whose elements are $(a_{.j}/n)^{1/2}$. It follows from (10.5) and (10.4) that, in this special case, $\mathbf{T}$ is of unit rank and its single latent root is unity.

In view of this last result we infer that the term '$-1$' in the general criterion (10.3) refers to a latent root of the matrix $\mathbf{T}$ arising solely from the expected values of the observations in $\mathbf{A}$ and that a non-zero value of the criterion $X^2$ must depend on the other latent roots of $\mathbf{T}$. Moreover, the value of $X^2$ will be partitioned into additive components corresponding to these other latent roots. To establish the number of degrees of freedom to attribute to each we argue as follows. The observed matrix $\mathbf{A}$ has $pq$ cells. In eliminating the effects of the expected values from the cells of $\mathbf{A}$, $(p+q-1)$ constants are fitted, that is one for each row and column of the table less one, since the rows and columns separately add up to $n$. The number of degrees of freedom on which $X^2$ is based then is

$$pq - (p+q-1) = (p-1)(q-1),$$

a result which is familiar. Eliminating the effects of the expected frequencies, as we have seen above, involves eliminating one latent root from $\mathbf{T}$, thereby reducing its rank by one. To eliminate a second latent root from $\mathbf{T}$ involves the fitting of a further $(p+q-3)$

constants, hence the second latent root of **T** is based on $(p+q-3)$ degrees of freedom. In a similar manner the third latent roots of **T** is based on $(p+q-5)$ degrees of freedom and so on, reducing the number of degrees of freedom by 2 for each succeeding root.

It can be shown that the largest latent root of **T** is unity. The corresponding latent vector is **a** (see 10.5) and its elements are known in advance. Since we have assumed that $p \geqslant q$, **T** is of order $q$ but its rank may be less than $q$ if its rows (or columns) are not linearly independent. The latent roots and corresponding latent vectors of **T** can be calculated in the usual way and, as we will shortly see, they assist in the interpretation of the overall test of association.

*Example* 10.1

The data in Table 10.1 show the ages at which systematic teaching of Phonics (letter sounds) was commenced for samples of children from England, Wales and Scotland respectively.

The problem is to see if there is an association between the ages at which the children commence their instruction and the country in which they live. Since the sample size is so large a formal test of significance is hardly necessary and the main contrasts between countries are immediately clear from an examination of the percentage frequencies for the columns of the table. In brief, they show that for Scotland and Wales about half the children are in the 'under 65' age group after which the percentages decrease with age, precipitiously after 71 months in the case of Scotland. For English children instruction begins on average at a later age than in Scotland or Wales, the highest percentage falling in the age group 61–74 months.

Table 10.1. Children in three countries classified by age for instruction in phonics

| *Age in months* | *England* | | *Wales* | | *Scotland* | | *Total* |
|---|---|---|---|---|---|---|---|
| | *n* | % | *n* | % | *n* | % | |
| Under 65 | 2 795 | (31) | 310 | (48) | 723 | (54) | 3 828 |
| 66–71 | 3 134 | (34) | 153 | (24) | 541 | (41) | 3 828 |
| 72–77 | 1 980 | (22) | 110 | (17) | 56 | (4) | 2 146 |
| 78–89 | 1 197 | (13) | 71 | (11) | 10 | (1) | 1 278 |
| Total | 9 106 | (100) | 644 | (100) | 1330 | (100) | 11 080 |

It is now of interest to see how an analysis of the data following the procedure outlined earlier reflects these general deductions. The **T**-matrix (10.2) was calculated and found to be

$$\mathbf{T} = \begin{bmatrix} 0.8296 & 0.2146 & 0.2965 \\ 0.2146 & 0.0634 & 0.0903 \\ 0.2945 & 0.0903 & 0.1613 \end{bmatrix}$$

Its trace is 1.0543 and, using (10.3), we find $X^2 = 601.64$ which, when referred to the chi-square distribution with 6 degrees of freedom, is found to be highly significant as anticipated. The latent roots of **T** are respectively

$$\lambda_1 = 1.0000, \qquad \lambda_2 = 0.0502, \qquad \lambda_3 = 0.0041,$$

The first is unity as expected, and is hereafter ignored. Each of the others may be tested for significance by the criterion

$$X^2 = n\lambda_j, (j > 1)$$

derived from (10.3). For $\lambda_2$ we find $X_2^2 = 556.22$, based on $(p + q - 3) = 4$ degrees of freedom, and for $\lambda_3$ we find $x_3^2 = 45.42$ based on 2 degrees of freedom (with $X^2 = X_1^2 + X_2^2$). Since both results are significant we conclude that **T** is of full rank, indicating that the frequencies in the columns of Table 10.1 are not related in a linear manner; on the contrary significant differences between them are now known to exist.

Next we turn to the latent vectors corresponding to $\lambda_2$ and $\lambda_3$. In normalized form, and written as rows, they are respectively:

$$\begin{array}{lllll} \mathbf{u}_2 & : & \{0.3912 & -0.1716 & -0.9042\} \\ \mathbf{u}_3 & : & \{0.1585 & -0.9552 & 0.2499\}. \end{array}$$

In order, the weights refer to the data for England, Wales and Scotland and inspection of the elements of $\mathbf{u}_2$ shows that the component $X_2^2 = 556.22$, which represents 92.5% of the overall criterion value, relates primarily to a contrast between the frequencies for England and Scotland. The fact that the weight for Wales has the same sign as that for Scotland brings out the partial resemblance between the two countries shown by the percentages for them in the first two rows of Table 10.1.

After the effect of the component $X_2^2$ has been removed from the data a lesser contrast still remains, corresponding to $X_3^2$. Inspection of the elements of $\mathbf{u}_3$ shows that it is one between Wales and the other two countries. It reflects the fact that, unlike England and

Scotland the percentages for Wales decrease fairly uniformly with increasing age. The main benefit of the analysis is its indication that the over-riding contrast in the data is that between the percentages for England and Scotland.

*Example* 10.2

Another useful feature of the Williams-type analysis is illustrated by the following example. In Table 10.2 a sample of 650 children from 3-child families (no brothers or sisters included), who were referred for psychiatric treatment, are classified according to psychiatric diagnosis of illness and ordinal position in the family.

The problem was to test for possible association between diagnosis and ordinal position.

For these data the **T**-matrix was found to be

$$\mathbf{T} = \begin{bmatrix} 0.34705 & 0.36167 & 0.30808 \\ 0.36167 & 0.37698 & 0.32013 \\ 0.30808 & 0.32013 & 0.28481 \end{bmatrix},$$

with latent roots

$$\lambda_1 = 1.0, \qquad \lambda_2 = 0.008838, \qquad \lambda_3 = 0.000001,$$

Since the third latent root is found to be (approximately) zero the matrix **T** may be considered not to be of full rank. This indicates that the independence of the rows (or columns) of Table 10.2 is in doubt; consequently the analysis immediately reveals an important feature of the data. In this simple example the 'cause' of the zero root is easy to detect for the ratios of corresponding elements in

Table 10.2 Children classified by diagnosis and ordinal position in family

| *Diagnosis* | *Ordinal position* | | | *Total* |
|---|---|---|---|---|
| | *1st* | *2nd* | *3rd* | |
| Anxiety reaction | 70 | 75 | 68 | 213 |
| Schizophrenic reaction | 57 | 61 | 52 | 170 |
| Depressive reaction | 98 | 108 | 61 | 267 |
| Total | 225 | 244 | 181 | 650 |

the first two rows of the table are

1.228, 1.230, 1.308,

respectively, which are roughly equal. Hence we may reasonably rule out the possibility of an association between *ordinal position* in the family and *anxiety* and *schizophrenic* reaction.

## Categories with a natural order

It is frequently the case that the $p$ rows and $q$ columns of a contingency table both have a natural order (see Maxwell, 1964, Chapter 4). Under these conditions the contingency table may be thought of as a bivariate frequency distribution for two continuous variates (the attributes) whose scales have been divided into $p$ and $q$ intervals respectively. The simplest case is the fourfold table ($p = q = 2$) and for it we have the well-known relationship

$$\phi^2 = X^2/n, \tag{10.6}$$

based on 1 degree of freedom, in which $\phi$ (representing the phi-coefficient) is an estimate of the product-moment correlation coefficient between the two variates under assumptions of normality. In the notation of this chapter (10.6) may be written as

$$\phi^2 = \text{tr.}\mathbf{T} - 1, \tag{10.7}$$

and the significance of $\phi$ would depend on the second latent root, $\lambda_2$, of $\mathbf{T}$.

For $q > 2$ a straightforward case is that in which the frequency distribution is bivariate normal. In this case, neglecting $\lambda_1$ as before, $\lambda_2$ will be the only non-zero latent root of $\mathbf{T}$ when association exists. Conversely, an approximate criterion for testing for non-linearity between two variates is

$$X^2 = n(\text{tr.}\mathbf{T} - \lambda_2 - 1), \tag{10.8}$$

which can be referred to the chi-square distribution with $(p-2)(q-2)$ degrees of freedom. In effect it is a test of the significance of the combined effects of the latent roots of $\mathbf{T}$ after the second, and a significant result would indicate departure from bivariate normality in the distribution of the data.

The latent vectors of $\mathbf{T}$ may be employed to find scores for the mid-points of the intervals into which the scales of the variates are divided. For each latent root there will be a separate set of scores for each attribute. For example, for $\lambda_j (j > 1)$ the corresponding vector $x_j$ will provide the scores for the attribute with the $q$ intervals. To find the scores for the other attribute we proceed as follows. We note that the matrix $\mathbf{T} = \mathbf{G}'\mathbf{G}$ is of order $q$ and that

$$(\mathbf{G}'\mathbf{G})\mathbf{x}_j = \lambda_j \mathbf{x}_j.$$

Premultiplying this equation by $\mathbf{G}$ we obtain,

$$(\mathbf{G}\mathbf{G}')\mathbf{G}\mathbf{x}_j = \lambda_j \mathbf{G}\mathbf{x}_j. \tag{10.9}$$

Now the matrix $\mathbf{G}\mathbf{G}'$ is of order $p$ and from (10.9) it follows that the column vector $\mathbf{G}\mathbf{x}_j$, of the same order, is the latent vector of it corresponding to the latent root $\lambda_j$. Hence the elements of $\mathbf{G}\mathbf{x}_j$ provide scores for the attribute which has $p$ intervals. Scaling of the elements of the two vectors is optional. Each vector could be normalized, and if in addition the normalized vectors are multiplied by $\lambda_j^{1/2}$ the scores obtained will reflect the magnitude of the latent root to which they refer. A further scaling procedure still is suggested by Williams (1952).

*Example* 10.3

A psychiatrist rated 387 patients on the two variables *worry* and *tension.* For each he used five categories ranging from one for patients who did not show the symptom up to a fifth category for patients who showed the symptom in an extreme form. The frequencies are given in Table 10.3 and the 63 patients in its S. W. corner are those who did not show either symptom.

Table 10.3 Contingency Table for the variates 'Worry' and 'Tension'

| | *Worry* | | | | | *Total* |
|---|---|---|---|---|---|---|
| Tension | 3 | 2 | 5 | 10 | 11 | 31 |
| | 11 | 8 | 16 | 35 | 19 | 89 |
| | 28 | 13 | 23 | 33 | 6 | 103 |
| | 27 | 11 | 23 | 12 | 5 | 78 |
| | 63 | 10 | 9 | 4 | 0 | 86 |
| Total | 132 | 44 | 76 | 94 | 41 | 387 |

The **T**-matrix for the data was calculated. Its second and third latent roots only were found to contribute significantly to the overall criterion value, the results being:

| | *Root* | $X^2$ | *d.f.* | *P* |
|---|---|---|---|---|
| $\lambda_2$ | 0.290482 | 112.42 | 7 | $< 0.001$ |
| $\lambda_3$ | 0.044833 | 17.35 | 5 | $< 0.01$ |

For $\lambda_2$, which is the dominant root affecting the criterion, the normalized vector of weights for *worry* (running from left to right with respect to Table 10.3) was found to be

0.704  0.074  – 0.082  – 0.462  – 0.528,

the corresponding vector for *tension* (running from bottom to top of Table 10.3) being

0.739  0.121  – 0.077  – 0.516  – 0.408.

Both sets of weights show a reasonably consistent trend and reflect the tendency towards bivariate normality in the table. For $\lambda_3$, which numerically is much smaller than $\lambda_2$, the corresponding normalized vectors are:

0.347  – 0.249  – 0.551  – 0.201  0.689, and
0.447  – 0.407  – 0.527  0.170  0.573.

Inspection of them suggests that, once the matrix **T** has had the effects of $\lambda_1$ (concerned with expected values) and $\lambda_2$ removed from it, the residuals which remain show a contrast between elements on the perimeters of the matrix with those more central to it. They reflect the truncated nature of the lateral distributions in Table 10.3 revealing, on the one hand, the known tendency amongst psychiatrists to rate a symptom as being absent if it viewed not to exist in a pathological sense. On the other hand, they suggest that the fifth category on each scale does not adequately allow for discrimination between patients who have the symptoms in a severe form.

### General comments

The methods of analysis described in this chapter cannot readily be extended to multidimensional contingency tables, which occur regularly in research work in the medical and social sciences. But other very adaptable methods, which involve the fitting of log-linear models, are available (see Birch, 1963; Bishop, 1969; and Goodman, 1971). They will be described in *Analysis of Contingency Tables* by B.S. Everitt which is derived from the present author's book *Analysing Qualitative Data*, now out of print.

CHAPTER ELEVEN

# Analysis of Variance in Matrix Notation

## Introduction

Research workers have at their disposal a wide variety of experimental designs which enable them to assess the relative effects of different factors affecting observations made in controlled investigations. In general these effects are represented by parameters in a linear model, estimates of which are required. Since the variety of designs is large the natural tendency, when writing computer programs, is to arrange matters so that a single program can cope with a whole class of designs, specific designs within the class being special cases. Some of the mathematical devices employed in the construction of such programs are discussed in this chapter.

## Experimental designs

At the outset we consider a straightforward two-way analysis of variance design involving two treatments, say $A$ and $B$. For simplicity assume that each is at only two levels, say $A_1$ and $A_2$ and $B_1$ and $B_2$. There are then 4 treatment combinations and the population means for observations on each may be displayed as in Table 11.1. In the table $\theta_{11}$ is the mean for the observations having the treatment combination $A_1B_1$, and so on; $\alpha_1$ is the mean for the observations involving $A_1$; $\beta_1$ is the mean for the observation involving $B_1$, and so on; while $\mu$ is the overall mean for all observations. If the observations are expressed as deviations from the overall mean then $\mu$ is set equal to zero and it follows that

$$\alpha_1 + \alpha_2 = 0; \beta_1 + \beta_2 = 0; \theta_{11} + \theta_{12} + \theta_{21} + \theta_{22} = 0. \quad (11.1)$$

The linear model for a two-way analysis of variance design with

Table 11.1 Mean scores for 4 treatment combinations

| | $B_1$ | $B_2$ | |
|---|---|---|---|
| $A_1$ | $\theta_{11}$ | $\theta_{12}$ | $\alpha_1$ |
| $A_2$ | $\theta_{21}$ | $\theta_{22}$ | $\alpha_2$ |
| | $\beta_1$ | $\beta_2$ | $\mu$ |

interaction is

$$y_{ijk} = \mu + \alpha_i + \beta_j + (\alpha\beta)_{ij} + e_{ijk}, \tag{11.2}$$

in which $y_{ijk}$ is the $k$th observation under the treatment combination $A_iB_j$, and $e_{ijk}$ is the corresponding 'error' term. Since (11.2) includes $\mu$ as an explicit term the other parameters in the model are subject to the constraints given in (11.1), the four interaction effects $(\alpha\beta)_{ij}$ having the values

$$(\alpha\beta)_{ij} = \theta_{ij} - \alpha_i - \beta_j. \tag{11.3}$$

In the model it is assumed that the error terms are normally distributed about zero with equal variance in the cells of Table 11.1.

Let $\boldsymbol{\beta}$ be a column vector with the parameters

$$\{\mu \;\; \alpha_1 \;\; \alpha_2 \;\; \beta_1 \;\; \beta_2 \;\; (\alpha\beta)_{11} \;\; (\alpha\beta)_{12} \;\; (\alpha\beta)_{21} \;\; (\alpha\beta)_{22}\} \tag{11.4}$$

as elements. Then the contribution of the parameters to each observation $y_{ijk}$ can be expressed as a vector product $\mathbf{a}\boldsymbol{\beta}$. For example, for the $k$th observation under the treatment combination $A_1B_1$, the row vector **a** is

$$[1 \quad 1 \quad 0 \quad 1 \quad 0 \quad 1 \quad 0 \quad 0 \quad 0]$$

and,

$$y_{11k} = \mu + \alpha_1 + \beta_1 + (\alpha\beta)_{11} + e_{11k}.$$

For $n$ observations under $A_1B_1$ there will be $n$ equations of this form differing only in their error terms. Similarly for $n$ observations under $A_1B_2$ the vector **a** is

$$[1 \quad 1 \quad 0 \quad 0 \quad 1 \quad 0 \quad 1 \quad 0 \quad 0],$$

and the $n$ equations will be

$$y_{12k} = \mu + \alpha_1 + \beta_2 + (\alpha\beta)_{12} + e_{12k}, \; (k = 1, \ldots, n),$$

and so on. All $4n$ **a**-vectors form a matrix **A** of order $4n$ by 9 and the model for the data as a whole can be written as

$$\mathbf{y} = \mathbf{A}\boldsymbol{\beta} + \mathbf{e}, \tag{11.5}$$

in which **y** is a column vector containing the $4n$ observation and **e** a similar vector containing the residuals. This equation has the same form as the regression equation (7.2) and to find a least square estimate **b** of $\boldsymbol{\beta}$ we may use (7.5), obtaining

$$\mathbf{b} = (\mathbf{A}'\mathbf{A})^{-1}\mathbf{A}'\mathbf{y}.$$

Unfortunately a difficulty now arises for, because of the constraints on the parameters (11.1), the matrix $(\mathbf{A}'\mathbf{A})$ is deficient in rank: its inverse does not exist and as a consequence all the parameters in $\boldsymbol{\beta}$ cannot be estimated simultaneously. The vector $\boldsymbol{\beta}$ is then said 'not to be estimable'.

In seeing how the difficulty is overcome let us assume for convenience that $n = 2$. The **A**-matrix then is

$$\mathbf{A} = \begin{bmatrix} 1 & 1 & 0 & 1 & 0 & 1 & 0 & 0 & 0 \\ 1 & 1 & 0 & 1 & 0 & 1 & 0 & 0 & 0 \\ 1 & 1 & 0 & 0 & 1 & 0 & 1 & 0 & 0 \\ 1 & 1 & 0 & 0 & 1 & 0 & 1 & 0 & 0 \\ 1 & 0 & 1 & 1 & 0 & 0 & 0 & 1 & 0 \\ 1 & 0 & 1 & 1 & 0 & 0 & 0 & 1 & 0 \\ 1 & 0 & 1 & 0 & 1 & 0 & 0 & 0 & 1 \\ 1 & 0 & 1 & 0 & 1 & 0 & 0 & 0 & 1 \end{bmatrix} \begin{matrix} \left.\vphantom{\begin{matrix}1\\1\end{matrix}}\right\} \text{for } A_1B_1 \\ \left.\vphantom{\begin{matrix}1\\1\end{matrix}}\right\} \text{for } A_1B_2 \\ \left.\vphantom{\begin{matrix}1\\1\end{matrix}}\right\} \text{for } A_2B_1 \\ \left.\vphantom{\begin{matrix}1\\1\end{matrix}}\right\} \text{for } A_2B_2 \end{matrix}$$

The deficiency of rank in the matrix is now obvious for columns 2 and 3 when added give column 1; so do columns 4 and 5, and also columns 6 to 9; moreover these combinations of columns correspond to the constraints shown in (11.1). We also note that in the matrix the elements of the last four columns are in turn the products of the elements in pairs of the columns 2 to 5 inclusive. For example column 6, which represents the interaction effect $(\alpha\beta)_{11}$, is the product of corresponding elements in columns 2 and 4, which represent respectively the main effects $\alpha_1$ and $\beta_1$, and so on.

Table 11.1 has 4 cells and after $\mu$ has been estimated only 3 degrees of freedom ramain for estimating the other eight parameters in the model. This fact, when taken in conjunction with the constraints given in (11.1), suggests that for example we might estimate the contrast $(\alpha_1 - \alpha_2)$, or equivalently $(\alpha_1 - \alpha_2)/2$, which involves only a single degree of freedom. Next we note that

$$(\alpha_1 - \alpha_2)/2 = \alpha_1 - (\alpha_1 + \alpha_2)/2 = \alpha_1, \tag{11.6}$$

hence an estimate of $(\alpha_1 - \alpha_2)/2$ will provide an estimate of $\alpha_1$, while it follows that $\alpha_2 = -\alpha_1$. Similarly, we can find estimates of $\beta_1$ and $(\alpha\beta)_{11}$ using the two remaining degrees of freedom, while estimates of the remaining parameters in (11.2) follow by deduction.

The suggested reparameterization of the model would affect the matrix **A**. Columns 2 and 3 in it would now be replaced by a single column, corresponding to the contrast $(\alpha_1 - \alpha_2)/2$, with elements

$$\{1 \quad 1 \quad 1 \quad 1 \quad -1 \quad -1 \quad -1 \quad -1\}.$$

Similarly, columns 4 and 5 in **A** would be replaced by a column with elements

$$\{1 \quad 1 \quad -1 \quad -1 \quad 1 \quad 1 \quad -1 \quad -1\},$$

corresponding to the contrast $(\beta_1 - \beta_2)/2$. The remaining 4 columns of **A**, representing interaction terms, would be replaced by a single column, namely

$$\{1 \quad 1 \quad -1 \quad -1 \quad -1 \quad -1 \quad 1 \quad 1\},$$

with elements given by the products of corresponding elements in the two earlier columns.

The new **A**-matrix, which we will denote by $\mathbf{A}_1$ with transpose $\mathbf{A}_1'$, now is

$$\mathbf{A}_1' = \begin{bmatrix} 1 & 1 & 1 & 1 & 1 & 1 & 1 & 1 \\ 1 & 1 & 1 & 1 & -1 & -1 & -1 & -1 \\ 1 & 1 & -1 & -1 & 1 & 1 & -1 & -1 \\ 1 & 1 & -1 & -1 & -1 & -1 & 1 & 1 \end{bmatrix}.$$

It is of rank four and the inverse of $\mathbf{A}_1'\mathbf{A}_1$ exists. The corresponding column vector of parameters $\boldsymbol{\beta}_1$ has elements

$$\{\mu \quad (\alpha_1 - \alpha_2)/2 \quad (\beta_1 - \beta_2)/2 \quad (\alpha_1 - \alpha_2)(\beta_1 - \beta_2)/4\}. \tag{11.7}$$

The model now becomes

$$\mathbf{y} = \mathbf{A}_1\boldsymbol{\beta}_1 + \mathbf{e}, \tag{11.8}$$

and the least square estimate $\mathbf{b}_1$ of $\boldsymbol{\beta}_1$ is given by

$$\mathbf{b}_1 = (\mathbf{A}_1'\mathbf{A}_1)^{-1}\mathbf{A}_1'\mathbf{y} \tag{11.9}$$

The estimate $\hat{\mathbf{y}}$ of $\mathbf{y}$, based on the parameters in the model, is

$$\hat{\mathbf{y}} = \mathbf{A}_1\mathbf{b}_1,$$

and the corresponding sum of squares is

$$\hat{\mathbf{y}}'\hat{\mathbf{y}} = \mathbf{b}_1'\mathbf{A}_1'\mathbf{A}_1\mathbf{b}_1. \tag{11.10}$$

The total uncorrected sum of square is $\mathbf{y}'\mathbf{y}$, hence the error sum of squares, by subtraction, is

$$\mathbf{e}'\mathbf{e} = \mathbf{y}'\mathbf{y} - \mathbf{b}_1'\mathbf{A}_1'\mathbf{A}_1\mathbf{b}_1.$$

Let $n_e$ denote the degrees of freedom for error, then the error variance, say $s^2$, is given by

$$s^2 = \mathbf{e}'\mathbf{e}/n_e, \tag{11.11}$$

As in the classic multiple regression model the error variances of the respective elements of $\mathbf{b}_1$ are the corresponding diagonal elements of $(\mathbf{A}_1'\mathbf{A}_1)^{-1}s^2$. Tests of significance of the contrasts in the model can now be carried out in the usual way and confidence intervals for the parameters can be derived.

The model too can readily be extended to cope with analysis of covariance. Suppose, for instance, that all the observations in the vector $\mathbf{y}$ are linearly affected by an extraneous variate, say the ages of the individuals involved, and that estimates of the parameters in the model are required with the effect of age partialled out. Then it is only necessary to add to the matrix $\mathbf{A}_1$ an additional column containing the scores for the individuals on the extraneous variate and proceed as before to estimate the parameters using the augmented $\mathbf{A}$-matrix. The model will now contain an additional parameter which will provide an estimate of the regression coefficient of the $\mathbf{y}$-scores on the extraneous variate, under the assumption that this coefficient is the same for each treatment combination. The significance of the regression coefficient can be tested in the same way as the estimates of the other parameters, while the latter will already be adjusted for the regression effect if such exists. Finally, by adding further columns to the matrix $\mathbf{A}_1$ the combined effect of several extraneous variates on the $\mathbf{y}$-scores can be taken into account simultaneously.

## Alternative procedures

Design matrices of the form $\mathbf{A}$, or $\mathbf{A}_1$, are clumsy to deal with. For instance if, in the previous example, $B$ had been at 3 levels then the matrix $\mathbf{A}_1$ would require six columns for effects of (say) the form

$\mu, (\alpha_1 - \alpha_2), (\beta_1 - \beta_3), (\beta_2 - \beta_3), (\alpha_1 - \alpha_2)(\beta_1 - \beta_3), (\alpha_1 - \alpha_2)(\beta_2 - \beta_3),$

and the matrix as a whole would be of order 6n by 6. Several devices have been proposed for simplifying the problem and some procedures in common use will now be illustrated for our $2 \times 2$ example.

Let **P** be a matrix composed of the last 4 columns of the matrix **A** (given above), that is a matrix composed of the columns representing the interaction effects in the model. Inspection of **P** shows that it is simply the *incidence* matrix for the observations, as the sums of its columns give the numbers of observations in the respective cells in the body of Table 11.1. The matrix **A** may now be expressed as a product $\mathbf{A} = \mathbf{PB}$ in which

$$\mathbf{B} = \begin{bmatrix} 1 & 1 & 0 & 1 & 0 & 1 & 0 & 0 & 0 \\ 1 & 1 & 0 & 0 & 1 & 0 & 1 & 0 & 0 \\ 1 & 0 & 1 & 1 & 0 & 0 & 0 & 1 & 0 \\ 1 & 0 & 1 & 0 & 1 & 0 & 0 & 0 & 1 \end{bmatrix},$$

its rows being the four distinct rows of **A**. Equation (11.5) may thus be written as

$$\mathbf{y} = (\mathbf{PB})\boldsymbol{\beta} + \mathbf{e}. \tag{11.12}$$

Next consider the matrix $\mathbf{A}_1$ or, more conveniently, its transpose $\mathbf{A}'_1$ given above. Inspection shows that the latter contains only four independent columns and when they are isolated and placed adjacent to each other we get the matrix (say **K**) given by

$$\mathbf{K} = \begin{bmatrix} 1 & 1 & 1 & 1 \\ 1 & 1 & -1 & -1 \\ 1 & -1 & 1 & -1 \\ 1 & -1 & -1 & 1 \end{bmatrix}.$$

This matrix is symmetric with rows (and columns) orthogonal and on inspection we find that

$$\mathbf{A}_1 = \mathbf{PK},$$

so that equation (11.8) may be written as

$$\mathbf{y} = \mathbf{PK}\boldsymbol{\beta}_1 + \mathbf{e}. \tag{11.13}$$

This is the basic equation for the analysis of the data.

In order to relate equation (11.13) to equation (11.12) let **L** be a matrix such that

$$\mathbf{L}\boldsymbol{\beta} = \boldsymbol{\beta}_1,$$

the elements of $\boldsymbol{\beta}$ and $\boldsymbol{\beta}_1$ being those given by (11.4) and (11.7) respectively. Equation (11.13) now becomes

$$\mathbf{y} = \mathbf{PKL}\boldsymbol{\beta} + \mathbf{e},$$

which on comparison with (11.12), gives $\mathbf{PB} = \mathbf{PKL}$, or $\mathbf{B} = \mathbf{KL}$, hence

$$\mathbf{L} = \mathbf{K}^{-1}\mathbf{B}.$$

In general the matrix $\mathbf{L}$ is not required, but for our example it is

$$\mathbf{L} = \begin{bmatrix} 1 & \frac{1}{2} & \frac{1}{2} & \frac{1}{2} & \frac{1}{2} & \frac{1}{4} & \frac{1}{4} & \frac{1}{4} & \frac{1}{4} \\ 0 & \frac{1}{2} & -\frac{1}{2} & 0 & 0 & \frac{1}{4} & \frac{1}{4} & \frac{1}{4} & \frac{1}{4} \\ 0 & 0 & 0 & \frac{1}{2} & -\frac{1}{2} & \frac{1}{4} & \frac{1}{4} & \frac{1}{4} & \frac{1}{4} \\ 0 & 0 & 0 & 0 & 0 & \frac{1}{4} & \frac{1}{4} & \frac{1}{4} & \frac{1}{4} \end{bmatrix},$$

and is such that the column vector $\boldsymbol{\beta}_1$ has elements as in (11.7), or in view of (11.6) and similar expressions, has elements

$$\{\mu \quad \alpha_1 \quad \beta_1 \quad (\alpha\beta)_{11}\}. \tag{11.14}$$

From the constraints it follows that $\alpha_1 = -\alpha_2, \beta_1 = -\beta_2$ and

$$(\alpha\beta)_{11} = -(\alpha\beta)_{12} = -(\alpha\beta)_{21} = (\alpha\beta)_{22},$$

so that estimates of all the parameters in (11.2) can be found.

In summary, we have

$$\begin{aligned} \mathbf{y} &= \mathbf{A}\boldsymbol{\beta} + \mathbf{e} \\ &= \mathbf{PB}\boldsymbol{\beta} + \mathbf{e} \\ &= \mathbf{PKL}\boldsymbol{\beta} + \mathbf{e} \\ &= \mathbf{PK}\boldsymbol{\beta}_1 + \mathbf{e}. \end{aligned} \tag{11.15}$$

From (11.15) the estimate $\hat{\mathbf{y}}$ of $\mathbf{y}$ is given by

$$\hat{\mathbf{y}} = \mathbf{PK}\mathbf{b}_1,$$

the estimate $\mathbf{b}_1$ of $\boldsymbol{\beta}_1$ being given by

$$\begin{aligned} \mathbf{b}_1 &= (\mathbf{K}'\mathbf{P}'\mathbf{PK})^{-1}\mathbf{K}'\mathbf{P}'\mathbf{y} \\ &= (\mathbf{K}'\mathbf{DK})^{-1}\mathbf{K}'\mathbf{P}'\mathbf{y}, \end{aligned} \tag{11.16}$$

in which $\mathbf{D} = \mathbf{P}'\mathbf{P}$ is a diagonal matrix whose elements are the numbers of observations in the respective cells of Table 11.1. Now $\mathbf{P}'\mathbf{y} = \mathbf{D}\boldsymbol{\theta}$, in which $\theta$ is a column vector with elements

$$\{\theta_{11} \quad \theta_{12} \quad \theta_{21} \quad \theta_{22}\},$$

hence (11.16) may be written simply as

$$\mathbf{b}_1 = (\mathbf{K}'\mathbf{D}\mathbf{K})^{-1}\mathbf{K}'\mathbf{D}\boldsymbol{\theta}, \tag{11.17}$$

and the matrices $\mathbf{A}$ and $\mathbf{A}_1$ have been completely dispensed with. In practice the estimate $\hat{\boldsymbol{\theta}}$ of $\boldsymbol{\theta}$, obtained from the data to hand, would be employed in (11.17).

*Example* 11.1

To illustrate some of the results given above we consider a two-way example, as in Table 11.1, but one in which the numbers of replications for the treatment combinations are unequal. The data are shown in Table 11.2.

The linear model assumed is (11.2) with parameter constraints as in (11.1). The column vector $\mathbf{y}$ has as its elements

$$\{13 \quad 15 \quad 14 \quad 17 \quad 18 \quad 13 \quad 2 \quad 12 \quad 14 \quad 9 \quad 13\}.$$

As before let $\boldsymbol{\beta}_1$ be the parameter vector with elements as in (11.14). The appropriate design matrix $\mathbf{A}_1$ (with transpose $\mathbf{A}_1'$) of rank 4 is

$$\mathbf{A}_1' = \begin{bmatrix} 1 & 1 & 1 & 1 & 1 & 1 & 1 & 1 & 1 & 1 & 1 \\ 1 & 1 & 1 & 1 & 1 & 1 & -1 & -1 & -1 & -1 & -1 \\ 1 & 1 & 1 & -1 & -1 & -1 & 1 & -1 & -1 & -1 & -1 \\ 1 & 1 & 1 & -1 & -1 & -1 & -1 & 1 & 1 & 1 & 1 \end{bmatrix}.$$

To find an estimate $\mathbf{b}_1$ of $\boldsymbol{\beta}_1$ we may employ (11.9) which requires us to calculate $\mathbf{A}_1'\mathbf{A}_1$, and its inverse $(\mathbf{A}_1'\mathbf{A}_1)^{-1}$, together with $\mathbf{A}'\mathbf{y}$.

Table 11.2 Two way design with unequal replications

| | $T_1$ | $T_2$ | *Total* |
|---|---|---|---|
| | 13 | 17 | |
| $G_1$ | 15 | 18 | |
| | 14 | 13 | 90 |
| | 2 | 12 | |
| $G_2$ | | 14 | |
| | | 9 | |
| | | 13 | 50 |
| Total | 44 | 96 | 140 |

We find

$$\mathbf{A}_1'\mathbf{A}_1 = \begin{bmatrix} 11 & 1 & -3 & 3 \\ 1 & 11 & 3 & -3 \\ -3 & 3 & 11 & 1 \\ 3 & -3 & 1 & 11 \end{bmatrix}, \quad \mathbf{A}'\mathbf{y} = \begin{bmatrix} 140 \\ 40 \\ -52 \\ 40 \end{bmatrix},$$

and

$$(\mathbf{A}_1'\mathbf{A}_1)^{-1} = \begin{bmatrix} 0.11979 & -0.03646 & 0.04688 & -0.04688 \\ -0.03646 & 0.11979 & -0.04688 & 0.04688 \\ 0.04688 & -0.04688 & 0.11979 & -0.03646 \\ -0.04688 & 0.04688 & -0.03646 & 0.11979 \end{bmatrix}$$

By (11.9) the column vector $\mathbf{b}_1$ of parameter estimates is found to be

$$\{11 \quad 4 \quad -3 \quad 2\}.$$

The same vector would have been found had we used (11.16) or (11.17). The total uncorrected sum of squares of the 11 observations is $\mathbf{y}'\mathbf{y} = 1966$, while the sum of squares accounted for by the parameters in the model, by (11.10), is found to be 1936. Hence the sum of squares for error is

$$\mathbf{e}'\mathbf{e} = 1966 - 1936 = 30,$$

and since it is based on 7 degrees of freedom the estimate of error variance is $s^2 = 30/7 = 4.2857$. The estimates of the error variances of the elements of $\mathbf{b}_1$ are the diagonal elements of $(\mathbf{A}_1'\mathbf{A}_1)^{-1}s^2$ and each is 0.5134. From $\mathbf{b}_1$ we have $\hat{\alpha}_1 = (\hat{\alpha}_1 - \hat{\alpha}_2)/2 = 4$, and it may be tested for significance by the $F$-ratio, $F = 4^2/0.5134 = 31.16$, with 1 and 7 degrees of freedom. Similar tests can be performed on the remaining elements of $\mathbf{b}_1$. Confidence limits for the parameters may also be set up; for example the 95% confidence limits for the interaction term are $2 \pm 2.36\ (0.5134)^{1/2}$, namely 3.69 and 0.31, where 2.36 is derived from the $t$-distribution with 7 degrees of freedom.

## Non-orthogonal designs

In the example just considered the error variance matrix is not diagonal. This will always be the case when the number of observations in the cells of a table such as Table 11.1 vary and, as a consequence, tests of significance, as illustrated in Example 11.1, will not be truly independent of each other. The matter is trivial when the numbers vary only slightly, but otherwise it may be more serious and further consideration of the problem is of interest.

Let $\mathbf{T}$ be a lower triangular matrix such that

$$\mathbf{TT}' = \mathbf{K}'\mathbf{DK} = \mathbf{K}'\mathbf{P}'\mathbf{PK},$$

and note that $(\mathbf{T}')^{-1}\mathbf{T}' = \mathbf{I}$. Expression (11.15) may now be written as

$$\begin{aligned}\mathbf{y} &= \mathbf{PK}(\mathbf{T}')^{-1}\mathbf{T}'\boldsymbol{\beta}_1 + \mathbf{e}\\ &= \mathbf{Z}\boldsymbol{\beta}_2 + \mathbf{e},\end{aligned} \tag{11.17}$$

in which $\mathbf{Z} = \mathbf{PK}(\mathbf{T}')^{-1}$ and $\boldsymbol{\beta}_2 = \mathbf{T}'\boldsymbol{\beta}_1$. It is easy to verify that $\mathbf{Z}'\mathbf{Z} = \mathbf{I}$, hence the vector $\boldsymbol{\beta}_2$ provides a set of contrasts on the parameters in the model which in a sense, elaborated below, are independent of each other. For these the error variances are given by

$$(\mathbf{Z}'\mathbf{Z})^{-1}s^2 = s^2\mathbf{I}, \tag{11.18}$$

in which $s^2$, as before, is the pooled 'within cells' estimate of error variance. By (11.17) we have

$$\hat{\mathbf{y}} = \mathbf{Zb}_2, \tag{11.19}$$

in which $\mathbf{b}_2$ is the estimate of $\boldsymbol{\beta}_2$ and is given directly by $\mathbf{T}^{-1}\mathbf{K}'\mathbf{P}'\mathbf{y}$ or $\mathbf{T}^{-1}\mathbf{K}'\mathbf{D}\hat{\boldsymbol{\theta}}$. We note that

$$\begin{aligned}\hat{\mathbf{y}}'\hat{\mathbf{y}} &= \mathbf{b}_2'\mathbf{Z}'\mathbf{Zb}_2\\ &= \mathbf{b}_2'\mathbf{b}_2\\ &= b_1^2 + b_2^2 + \ldots + b_t^2\end{aligned} \tag{11.20}$$

in which $t$ is the rank of the matrix $\mathbf{K}$. Hence the sum of squares due to the parameters in the model has been decomposed into $t$ independent additive components. In interpreting (11.20) let us look again at our example.

Example 11.1 (continued)

With $\mathbf{K}$ and $\mathbf{D}$ as before, we find

$$\mathbf{K}'\mathbf{DK} = \begin{bmatrix} 11 & 1 & -3 & 3\\ 1 & 11 & 3 & -3\\ -3 & 3 & 11 & 1\\ 3 & -3 & 1 & 11\end{bmatrix},$$

$$\mathbf{T} = \begin{bmatrix} 3.3166 & & & \\ 0.3015 & 3.3029 & & \\ -0.9045 & 0.9909 & 3.0332 & \\ 0.9045 & -0.9909 & 0.9231 & 2.8893\end{bmatrix},$$

where $\mathbf{TT}' = \mathbf{K}'\mathbf{DK}$. An estimate $\mathbf{b}_2$ of $\boldsymbol{\beta}_2$ is then given by $\mathbf{b}_2 = \mathbf{T}'\mathbf{b}_1$, where $\mathbf{b}_1$ is the column vector of parameter estimates already found, namely,

$$\{11 \quad 4 \quad -3 \quad 2\}.$$

The vector $\mathbf{b}_2$ has elements

$$\{42.2116 \quad 8.2572 \quad -7.2532 \quad 5.7785\},$$

and, using (11.20), we find the sums of squares due to the parameters in the model to be 1936, as in the earlier analysis.

Now consider the elements of $\mathbf{b}_2$ in turn. The square of the first, namely $42.2116^2 = 1781.82$, is the correction term, namely $140^2/11 = 1781.82$, in the analysis of variance of the data. The square of the second, namely $9.2572^2 = 68.18$, is the sum of squares due to the parameters $\alpha_1$ and $\alpha_2$ if the model

$$y_{ik} = \mu + \alpha_i + e_{ik},$$

is fitted to the data, thus ignoring the effect of the factor $B$ and the interaction between factors $A$ and $B$. To check this we use the first two columns only of the original matrix $\mathbf{A}_1$. We then have

$$(\mathbf{A}'\mathbf{A}) = \begin{bmatrix} 11 & 1 \\ 1 & 11 \end{bmatrix},$$

and with appropriate substitution in expressions (11.8) and (11.9) we find the corresponding vector of parameters to be $\{12.5 \quad 2.5\}$ and the sum of squares accounted for by them to be 1850. But since the 'correction' sum of square is 1781.82 the sum of squares due to the $\alpha$'s is

$$1850 - 1781.82 = 68.18,$$

as indicated above.

In a similar manner if we fit the model

$$y_{ijk} = \mu + \alpha_i + \beta_j + e_{ijk}$$

to the data, thus ignoring the interaction term, the vector of parameter estimates is found to be

$$\{11.783 \quad 3.217 \quad -2.391\}$$

and the sum of squares accounted for by the parameters to be 1902.61. If from the latter we subtract the correction term and the sum of squares due to fitting the $\alpha$'s alone, we get

$$1902.61 - 1781.82 - 68.18 = 52.63,$$

which is the square of the third element, namely 7.2532, in $\mathbf{b}_2$. The additional sum of the squares accounted for by including $\beta_j$, in addition to $\alpha_i$, in the model thus is 52.63. In summary, the successive terms in (11.20) refer to the additional sum of squares accounted for as additional parameters are added to the model. If at a given stage in the process we wish to apply tests of significance then the estimate of 'error' variance will be based on the residual sum of squares and the residual degrees of freedom at that stage. When a break-down of sums of squares, as provided by (11.20), is contemplated in an analysis the order in which the parameters are placed in a linear equation such as (11.2) matters but a choice, based on prior knowledge of the relative importance of different factors, is often possible.

## Generalization

From the previous discussion it is clear that the main problem which arises, in the use of the mathematical devices described in this chapter, is that of generating an appropriate **K**-matrix. But this can readily be overcome, as the matrix required can be derived from a series of simple contrast (basis) matrices one for each of the factors involved in a particular analysis.

For the simple example already discussed, involving two factors $A$ and $B$ each at two levels, the two contrast matrices are each of the form

$$\begin{bmatrix} 1 & 1 \\ 1 & -1 \end{bmatrix}.$$

For the $A$-factor the first column of this matrix is concerned with the estimate of $\mu$ and the second with the estimate of $\alpha_1$. Similarly, for the $B$-factor, the first column is concerned with the estimate of $\mu$ and the second with the estimate of $\beta_1$. To derive the appropriate **K**-matrix we take the direct product, or Kronecker product, of these two simple matrices.

For two matrices **A** and **B**, of order $p \times q$ and $m \times n$ respectively, the direct product is defined by

$$\mathbf{A}^*\mathbf{B} = \begin{bmatrix} a_{11}B & \dots & a_{1q}B \\ \cdot & & \cdot \\ \cdot & & \cdot \\ \cdot & & \cdot \\ a_{p1}B & & a_{pq}B \end{bmatrix}$$

Hence if both **A** and **B** are

$$\begin{bmatrix} 1 & 1 \\ 1 & -1 \end{bmatrix},$$

then

$$\mathbf{A}^*\mathbf{B} = \begin{bmatrix} 1 & 1 & 1 & 1 \\ 1 & -1 & 1 & -1 \\ 1 & 1 & -1 & -1 \\ 1 & -1 & -1 & 1 \end{bmatrix}.$$

The order of the multiplication defines the order of the columns in the product matrix with respect to the parameters. In **A*B** the columns refer in turn to $\mu$, $\beta_1$, $\alpha_1$ and $(\alpha\beta)_{11}$ respectively. In other words the matrix **A*B** is the matrix **K** found earlier with its second and third columns interchanged, but once this has been noted the matter is of trivial importance. To keep track of order it is helpful to look at the direct product of **A** and **B** taking the columns of **A** in turn. For example, if its first column is denoted by $a_1$, we have

$$\mathbf{a}_1^*\mathbf{B} = \begin{bmatrix} 1 \\ 1 \end{bmatrix} \begin{bmatrix} 1 & 1 \\ 1 & -1 \end{bmatrix} = \begin{bmatrix} 1 & 1 \\ 1 & -1 \\ 1 & 1 \\ 1 & -1 \end{bmatrix},$$

and it is clear that the second column in the product matrix refers to $\beta_1$ since the second column in **B** refers to this parameter.

As a further example consider a design in which the factor $A$ is at two levels but the factor $B$ is at three levels. The respective simple contrast matrices of full rank then are

$$\mathbf{A} = \begin{bmatrix} 1 & 1 \\ 1 & -1 \end{bmatrix} \quad \text{and } \mathbf{B} = \begin{bmatrix} 1 & 1 & 0 \\ 1 & 0 & 1 \\ 1 & -1 & -1 \end{bmatrix},$$

the second and third columns of **B** being chosen to provide simultaneously estimates of $\beta_1$ and $\beta_2$ based on the contrasts $(\beta_1 - \beta_3)$ and $(\beta_2 - \beta_3)$ respectively, with $\beta_1 + \beta_2 + \beta_3 = 0$. The direct product of **A** and **B** gives an appropriate **K**-matrix for the design, namely

$$\mathbf{A}^*\mathbf{B} = \begin{bmatrix} 1 & 1 & 0 & 1 & 1 & 0 \\ 1 & 0 & 1 & 1 & 0 & 1 \\ 1 & -1 & -1 & 1 & -1 & -1 \\ 1 & 1 & 0 & -1 & -1 & 0 \\ 1 & 0 & 1 & -1 & 0 & -1 \\ 1 & -1 & -1 & -1 & 1 & 1 \end{bmatrix},$$

in which the columns refer in turn to the parameters in the order

$$\mu, \beta_1, \beta_2, \alpha_1, (\alpha\beta)_{11}, (\alpha\beta)_{12}.$$

For more general and systematic procedures see Nelder (1965).

CHAPTER TWELVE

# Multivariate Analysis of Variance (MANOVA)

## Introduction

The methods described in Chapter 11 can readily be extended to the multivariate case. To introduce the discussion let us assume that in Table 11.1 $p$ variates, rather than just a single variate, are measured under each of the four treatment combinations. Each of the (scalar) parameters in the table must now be replaced by a vector of order $p$. For example $\alpha_1$ and $\alpha_2$ will become row vectors

$$\boldsymbol{\alpha}_1 = \{\alpha_{11}\ \alpha_{12} \dots \alpha_{1p}\}$$

and

$$\boldsymbol{\alpha}_2 = \{\alpha_{21}\ \alpha_{22} \dots \alpha_{2p}\}$$

and a test of the hypothesis that the two levels of treatment, $A_1$ and $A_2$, are equal will become a test of equality of corresponding pairs of elements in $\boldsymbol{\alpha}_1$ and $\boldsymbol{\alpha}_2$, that is a test that simultaneously

$$\begin{aligned} \alpha_{11} &= \alpha_{21}, \\ \alpha_{12} &= \alpha_{22}, \\ &\vdots \\ \alpha_{1p} &= \alpha_{2p}. \end{aligned}$$

The multivariate test has the advantage over $p$ univariate tests in that it takes possible correlation and differences in variance between the $p$ variates into account. The variates need not be in the same metric, but it is assumed that they have equal covariance matrices under each separate treatment combination.

## The model

For each of the $p$ variates separately a linear model similar to

(11.2) is assumed. In equation (11.5) the matrix **A** remains the same as in the univariate case but the column vector **y** is replaced by a matrix **Y** having a separate column for each of the $p$ variates. Similarly the column vector $\boldsymbol{\beta}$ of parameters is replaced by a matrix **B** having $p$ columns, and the single vector **e** is replaced by a matrix **E** with $p$ columns of error terms.

The reparameterization is the same as in the univariate case, and minimization of the trace of $\mathbf{E}'\mathbf{E}$ leads to the estimation equation

$$\hat{\mathbf{B}}_1 = (\mathbf{A}_1'\mathbf{A}_1)^{-1}\mathbf{A}_1'\mathbf{Y}, \tag{12.1}$$

corresponding to (11.9), each of the $p$ columns of **B** giving estimates for their respective variates of parameters as in (11.7). Following the procedure of Chapter 11 we also obtain the equation

$$\hat{\mathbf{B}}_1 = (\mathbf{K}'\mathbf{D}\mathbf{K})^{-1}\mathbf{K}'\mathbf{P}'\mathbf{Y}, \tag{12.2}$$

corresponding to (11.16), and the equation

$$\hat{\mathbf{B}}_1 = (\mathbf{K}'\mathbf{D}\mathbf{K})^{-1}\mathbf{K}'\mathbf{D}\boldsymbol{\Theta},$$

corresponding to (11.17), in which $\boldsymbol{\Theta}$ is a matrix whose rows are the mean vectors $\boldsymbol{\theta}$ of the variates, one for each of the treatment combinations.

The equation for estimating **Y** in the multivariate model is

$$\hat{\mathbf{Y}} = \mathbf{A}_1\hat{\mathbf{B}}_1$$

and the matrix of uncorrected sums of squares and cross-products due to estimation is

$$\begin{aligned}\hat{\mathbf{Y}}'\hat{\mathbf{Y}} &= \hat{\mathbf{B}}_1'\mathbf{A}_1'\mathbf{A}_1\hat{\mathbf{B}}_1 \\ &= \boldsymbol{\Theta}'\mathbf{D}\boldsymbol{\Theta},\end{aligned} \tag{12.4}$$

using (12.3). The corresponding matrix for the observed scores is $\mathbf{Y}'\mathbf{Y}$, hence the residual or 'within subclasses' matrix is

$$\mathbf{E}'\mathbf{E} = \mathbf{Y}'\mathbf{Y} - \boldsymbol{\Theta}'\mathbf{D}\boldsymbol{\Theta} \tag{12.5}$$

If $n_e$ represents the residual degrees of freedom the residual covariance matrix, say $\Sigma$, is estimated by

$$\hat{\boldsymbol{\Sigma}} = (\mathbf{Y}'\mathbf{Y} - \boldsymbol{\Theta}'\mathbf{D}\boldsymbol{\Theta})/n_e. \tag{12.6}$$

*Example* 12.1

As an illustration of multivariate analysis of variance we shall

Table 12.1 Scores on three variates in a two-way design

| | $T_1$ | | | $T_2$ | | |
|---|---|---|---|---|---|---|
| | $y_1$ | $y_2$ | $y_3$ | $y_1$ | $y_2$ | $y_3$ |
| $G_1$ | 13 | 20 | 17 | 17 | 16 | 22 |
| | 15 | 16 | 16 | 18 | 18 | 22 |
| | 14 | 21 | 15 | 13 | 17 | 16 |
| | (14) | (19) | (16) | (16) | (17) | (20) |
| $G_2$ | 2 | 3 | 8 | 12 | 14 | 16 |
| | | | | 14 | 14 | 19 |
| | | | | 9 | 10 | 13 |
| | | | | 13 | 14 | 16 |
| | (2) | (3) | (8) | (12) | (13) | (16) |

take an extension of Example 11.1. Denote the variate on which the scores in Table 11.2 were obtained by $y_1$ and assume that two further variates, $y_2$ and $y_3$, were observed, the scores on all three variates being those shown in Table 12.1.

The diagonal matrix **D** of subclass frequencies has elements

$$3 \quad 3 \quad 1 \quad 4,$$

respectively. The numbers in brackets in the table are the subclass means, hence the estimate $\hat{\boldsymbol{\Theta}}$ of $\boldsymbol{\Theta}$ is

$$\hat{\boldsymbol{\Theta}} = \begin{bmatrix} 14 & 19 & 16 \\ 16 & 17 & 20 \\ 2 & 3 & 8 \\ 12 & 13 & 16 \end{bmatrix}.$$

As there are 11 observations on each of the three variates the matrix **Y** is of order 11 × 3, its first column being that shown in Example 11.1 with corresponding columns for $y_2$ and $y_3$. We find

$$\mathbf{Y'Y} = \begin{bmatrix} 1966 & \text{(symmetric)} & \\ 2253 & 2663 & \\ 2448 & 2799 & 3100 \end{bmatrix}, \quad \boldsymbol{\Theta}'\mathbf{D}\boldsymbol{\Theta} = \begin{bmatrix} 1936 & \text{(symmetric)} & \\ 2240 & 2635 & \\ 2416 & 2788 & 3056 \end{bmatrix}.$$

Hence the residual matrix, by (11.5), is

$$\mathbf{E}'\mathbf{E} = \begin{bmatrix} 30 & \text{(symmetric)} & \\ 9 & 28 & \\ 32 & 11 & 44 \end{bmatrix}.$$

For this two-way analysis an appropriate **K**-matrix, as seen in Chapter 11, is

$$\mathbf{K} = \begin{bmatrix} 1 & 1 & 1 & 1 \\ 1 & 1 & -1 & -1 \\ 1 & -1 & 1 & -1 \\ 1 & -1 & -1 & 1 \end{bmatrix},$$

which for each of the three variates estimates the parameters in the order

$$\mu, \qquad \alpha_1, \qquad \beta_1, \qquad (\alpha\beta)_{11}.$$

Using (12.3) we find

$$\hat{\mathbf{B}}_1 = \begin{bmatrix} 11 & 13 & 15 \\ 4 & 5 & 3 \\ -3 & -2 & -3 \\ 2 & 3 & 1 \end{bmatrix}.$$

The entries in its first column are the estimates of the parameters found in Example 11.1 and correspond to the variate $y_1$. The entries in its other two columns are the parameter estimates for $y_2$ and $y_3$ respectively. In passing it is of interest to note that the parameter values given in $\hat{\mathbf{B}}_1$ are the values used when constructing the artificial data shown in Table 12.1 (random errors with zero means being then added to the scores on each variate in each subclass) and that expression (12.1) reproduces them exactly despite the unequal numbers of observations in the subclasses. In other words disproportionate subclass frequencies do not cause bias in the parameter estimates when the model fitted is appropriate.

Univariate tests of significance of the parameter estimates provided by $\hat{\mathbf{B}}_1$ can now be carried out as indicated in Example 11.1. To illustrate multivariate tests the analysis of the data must be carried a step further.

## Partitioning the matrix $\mathbf{\Theta}'\mathbf{D}\mathbf{\Theta}$

The matrix of sums of squares and cross-products due to the parameters is partitioned to show the contributions made to it by the several vectors of parameters in the model. The method given below is general in the sense that it applies equally to orthogonal and non-orthogonal designs. In the former case the elements of the diagonal matrix $\mathbf{D}$ of subclass frequencies are all equal and the partitioning gives estimates of main and interaction effects which are mutually independent of each other. Otherwise the partitioning corresponds to the order in which the parameters are estimated, as in the second analysis of the data in Example 11.1.

To partition $\mathbf{\Theta}'\mathbf{D}\mathbf{\Theta}$ we first factorise the matrix $\mathbf{K}'\mathbf{D}\mathbf{K}$ into the form $\mathbf{T}\mathbf{T}'$, where $\mathbf{T}$ is a lower triangular matrix and $\mathbf{T}'$ its transpose. The calculations are already given on p. 124. We then employ equation (11.19), adjusted to the multivariate case, that is

$$\hat{\mathbf{Y}} = \mathbf{Z}\hat{\mathbf{B}}_2, \tag{12.7}$$

in which

$$\mathbf{Z} = \mathbf{P}\mathbf{K}(\mathbf{T}')^{-1},$$

as before, while

$$\begin{aligned}\hat{\mathbf{B}}_2 &= \mathbf{T}'\hat{\mathbf{B}}_1 \\ &= \mathbf{T}^{-1}\mathbf{K}'\mathbf{D}\hat{\mathbf{\Theta}}.\end{aligned} \tag{12.8}$$

For our example we find

$$\hat{\mathbf{B}}_2 = \begin{bmatrix} 42.2116 & 49.1463 & 54.2720 \\ 8.2572 & 11.5601 & 5.9452 \\ -7.2532 & -3.2969 & -8.1763 \\ 5.7785 & 8.6677 & 2.8892 \end{bmatrix} \begin{matrix} = \mathbf{u}_1' \\ = \mathbf{u}_2' \\ = \mathbf{u}_3' \\ = \mathbf{u}_4', \end{matrix}$$

Denote its rows by the vectors $\mathbf{u}_1'$ to $\mathbf{u}_4'$ respectively. Note that the first column of $\hat{\mathbf{B}}_2$ is the vector $\mathbf{b}_2$ of page 125 and it refers to $y_1$, its other two columns refer to $y_2$ and $y_3$ respectively. Corresponding to equation (11.20) the partitioning of the matrix $\hat{\mathbf{Y}}'\hat{\mathbf{Y}} = \hat{\mathbf{\Theta}}'\mathbf{D}\mathbf{\Theta}$, is given by the equation

$$\begin{aligned}\hat{\mathbf{\Theta}}'\mathbf{D}\mathbf{\Theta} &= \mathbf{u}_1\mathbf{u}_1' + \mathbf{u}_2\mathbf{u}_2' + \mathbf{u}_3\mathbf{u}_3' + \mathbf{u}_4\mathbf{u}_4', \\ &= \mathbf{W}_\mu + \mathbf{W}_\alpha + \mathbf{W}_\beta + \mathbf{W}_{\alpha\beta} \quad \text{(say)},\end{aligned} \tag{12.9}$$

in which the $\mathbf{W}$'s are square symmetric matrices. Their numerical values are shown in Table 12.2.

### Tests of significance in MANOVA

Several criteria have been proposed for testing the significance

Table 12.2 The Partitioning of $\mathbf{\Theta}'\mathbf{D}\mathbf{\Theta}$

| *Source* | *S.S. and S.P.* | *d.f.* |
|---|---|---|
| Correction terms | $\begin{bmatrix} 1781.81 & \text{(symmetric)} & \\ 2074.54 & 2415.36 & \\ 2209.90 & 2667.27 & 2945.45 \end{bmatrix} = \mathbf{W}_\mu$ | 1 |
| *G*-effects, ignoring *T* and *GT* | $\begin{bmatrix} 68.18 & \text{(symmetric)} & \\ 95.54 & 133.64 & \\ 49.09 & 68.73 & 35.35 \end{bmatrix} = \mathbf{W}_\alpha$ | 1 |
| *T*-effects; *G* eliminated and *GT* ignored | $\begin{bmatrix} 52.62 & \text{(symmetric)} & \\ 23.91 & 10.87 & \\ 59.30 & 26.96 & 66.85 \end{bmatrix} = \mathbf{W}_\beta$ | 1 |
| *GT* effects, *G* and *T* eliminated | $\begin{bmatrix} 33.39 & \text{(symmetric)} & \\ 50.05 & 75.13 & \\ 16.70 & 25.04 & 8.35 \end{bmatrix} = \mathbf{W}_{\alpha\beta}$ | 1 |
| Total | $\begin{bmatrix} 1936.00 & \text{(symmetric)} & \\ 2244.00 & 2635.00 & \\ 2416.00 & 2788.00 & 3056.00 \end{bmatrix} = \hat{\mathbf{\Theta}}'\mathbf{D}\hat{\mathbf{\Theta}}$ | 4 |

of the separate effects in a multivariate analysis of variance and special statistical distributions are generally required. But standard computer programs carry out the relevant tests automatically. However, when a **W**-matrix (e.g. Table 12.2) is based on a single degree of freedom a simple $F$-ratio test is available (see Morrison 1967, p. 167). It is based on the single latent root, $\lambda$, of the matrix $\mathbf{W}\mathbf{R}^{-1}$ where $\mathbf{W}$ is an effects matrix based on the single degree of freedom and $\mathbf{R}$ is the residual matrix given by (12.5). The criterion then is

$$F = \{(n+1)/(m+1)\}\lambda, \tag{12.10}$$

with degrees of freedom $2m+2$ and $2n+2$, where

$$m = \tfrac{1}{2}\{|1-t|-1\},$$
$$n = \tfrac{1}{2}(N-t-p-1),$$

$N$ being the total sample size, $t$ the rank of the matrix $\mathbf{K}$, and $p$ the number of variates.

For our example

$$\mathbf{R}^{-1} = \begin{bmatrix} 0.1494 & -0.0059 & -0.1072 \\ -0.0059 & 0.0398 & -0.0057 \\ -0.1072 & -0.0057 & 0.1022 \end{bmatrix}.$$

To test for a significance interaction effect in Table 12.2 we find the product matrix

$$\mathbf{W}_{\alpha\beta}\mathbf{R}^{-1} = \begin{matrix} 2.903 & 1.701 & -2.158 \\ 4.356 & 2.552 & -3.239 \\ 1.452 & 0.850 & -1.080 \end{matrix} .$$

This matrix is of unit rank with trace equal to 4.375, which thus is the value of $\lambda$. With $N = 11, t = 4$ and $p = 3$ we find $m = 1$ and $n = 1.5$. Substituting these values in (12.10) gives $F = 5.47$ based on 4 and 5 degrees of freedom. Hence the interaction term is significant beyond the 5% level. The other effects in Table 12.2 can be tested in a similar manner. Univariate tests of significance can also be carried out on the data in Table 12.2 using the diagonal elements of the **W**-matrices and the corresponding elements of $R$. For example, to test the interaction term for $y_3$ the error variance is $44/7 = 6.286$ and the $F$-ratio is

$$F = 8.35/6.286 = 1.328,$$

which, with 1 and 7 degrees of freedom, is far from being significant. The inference is that the significant result found for the multivariate test for interaction is due to either $y_1$ or $y_2$, or both.

CHAPTER THIRTEEN

# Cluster Analysis and Miscellaneous Techniques

## Introduction

The techniques of multivariate analysis described in previous chapters all make use of some relatively well-defined underlying model. Linear relationships between the variates have in general been assumed and the populations employed were known in advance. In this chapter multivariate methods will be discussed, which are useful for an initial screening of data, for which little or no prior information is available and few assumptions can be made. They fall loosely under two headings, *cluster analysis* and *visual representations*. In general they should be regarded as descriptive *statistical* methods in the sense that they have no associated distribution theory or significance tests and so are unable to relate from sample to population. Indeed many of them treat the data to hand as the population.

## Cluster analysis techniques

The term *cluster analysis* embraces a loosely structured body of *ad hoc* algorithms, which are used in the exploration of data that arise from the measurement of a number of characteristics for each of an assorted collection of individuals or objects. The aim of the exploration is to see if the latter can be subdivided into groups or clusters which on the basis of the measurements can be shown to be relatively distinct or to belong together. The aim is clearly different from that of discriminant function analysis, or similar *assignment* techniques, used for the allocation of people to known groups. Unlike these, cluster analysis is concerned with the more difficult and intrinsically more interesting problem of discovering the groups in the first instance.

But cluster analysis has its own peculiar problems. The most serious of these is the lack of a satisfactory definition of exactly what constitutes a cluster. Because of this, most clustering techniques cannot be formulated in precise mathematical terms as can techniques such as confirmatory factor analysis (Chapter 6). The problem of definition has been considered by several authors, for example by Ling (1972), and is obviously crucial in many respects. However, since the main aim of this section is mearely to acquaint readers with some particular clustering techniques and their associated problems, the matter of definition will conveniently be ignored, and the word *cluster* or *group* used in an intuitive sense for a collection of 'similar' individuals or objects.

**Similarity and distance measures**

The first stage in many methods of cluster analysis is the conversion of the $(n \times p)$ matrix of data, $\mathbf{X}$, into an $(n \times n)$ matrix of inter-individual *similarities* or *dissimilarities*. The latter are terms for measures of the relationships between pairs of individuals, given the values of a set of $p$-variates for each. High similarity values indicate that the two individuals are alike with respect to the set of variates, whilst high dissimilarity indicates the opposite. Similarity measures generally take values between zero and one, whilst dissimilarity measures may take any positive value. A more fundamental difference between similarity and dissimilarity measures is introduced by requiring the latter to satisfy the *metric inequality*, (see Everitt, 1974, p. 56) in which case they are generally termed *distance* measures. Commonly used similarity coefficients are the product moment correlation coefficient and the simple matching coefficient for binary data. The best known distance measure is, of course, Euclidean. In some situations, for example, in psychological 'preference' experiments the data may actually arise as a similarity or dissimilarity matrix which might then be used as direct input into some clustering technique. Full accounts of similarity and dissimilarity measures are given in Sneath and Sokal, (1973), and by the author in his monograph.

**Common techniques**

Comprehensive reviews of clustering techniques are given by Cormack (1971) and by Everitt (1974), who lists the computer programs available. In this section a few of the most commonly used methods will be described briefly.

### (*a*) *Agglomerative hierarchical techniques*

The basic procedure with all methods of this type is similar; beginning with an inter-individual similarity or dissimilarity matrix they proceed by a series of successive fusions of the $n$ individuals into groups culminating at the stage where all individuals are in one group. At any particular stage in the procedure the methods fuse together the two individuals or two groups of individuals which are most similar. Differences between methods arise because of the differing ways of defining similarity between an individual and a group containing several individuals or between two such groups. For example, a method known as *group average clustering* defines similarity between clusters as the average similarity of all pairs of individuals in the two clusters. Another well known method of this type, namely *single-linkage clustering* defines intercluster similarity as the similarity between the two closest members of each.

Various difficulties are encountered when using these techniques in practice. In the first place they give no clear indication as to the 'correct' number of groups to consider for a given set of data (see Everitt, 1974, p. 59). A second difficulty is that the clustering achieved by a particular technique is likely to reflect any implicit assumptions it contains about the type of structure present in the data. A common assumption is that clusters, if any exist, will be spherical in shape: but if in fact they are not spherical, but (say) banana-shaped, spurious solutions may well be obtained (see Everitt, 1974, Figures 4.4 and 4.5).

A further problem is one discussed by Jardine and Sibson (1968). They point out that similarity and dissimilarity measures rarely have strict numerical significance. Because of the arbitrariness involved in the scaling of variables and in their fusion there is rarely any prior justification for using particular numerical values rather than values obtained by some monotonic transformation of them, such as their logarithms or square roots. In most cases the values should be regarded as having only *ordinal* significance. Jardine and Sibson point out that the only agglomerative hierarchical technique applicable in this case is single-linkage clustering, since the solutions produced by this method are invariant under monotonic transformations of the similarity measure. This has lead them to recommend this technique as the only mathematically acceptable one in the agglomerative class, a recommendation which has been criticised by other workers, for example, Williams *et al.* (1971).

### (*b*) *Optimization techniques*

A further class of clustering methods operate by forming clusters so as to optimize (minimize or maximize) a pre-defined *clustering criterion*, this being some numerical measure defined for every partition of the $n$ objects into say, $g$ groups, where $g$ has been fixed by the experimenter. An example of a commonly used criterion is minimization of the pooled within clusters sum of squares (see 9.2). Other criteria are discussed by the author in his monograph. Again with these techniques, as with those discussed in (a), there seem to be no acceptable methods for deciding on a particular value of $g$. In practice a range of values is used. A further problem arises in the actual optimization process. For any clustering criterion, say $C$, the obvious way to optimize is to consider every possible partition of the $n$ objects into $g$ groups, compute $C$ for each, and select that solution with the optimal value of $C$. However, because of the enormous number of partitions possible even with moderately sized $n$, this is not practicable and various *hill-climbing* (hill-descending in the case of minimization) algorithms are used in an attempt to reach the optimal partition. Starting with some initial partition of the $n$ objects into $g$ groups, each individual is considered for a possible move from the group he is in to one of the other groups, the move being made only if it causes an increase (decrease in the case of minimization) in the value of the clustering criterion. This process is continued until no move of a single individual causes improvement in the criterion value. However there is no guarantee that this method will converge on the optimal rather than on some sub-optimal partition. In practice attempts are made to overcome this problem of sub-optimal solutions by repeating the process with different initial partitions.

### (*c*) *Clustering by fitting mixtures of multivariate normal distributions*

Wolfe in a series of papers (1967; 1969; 1970), describes a clustering technique based on fitting mixtures of multivariate normal distributions. In essence he assumes that the data are a sample from some population in which the distribution of the variates has the form $f(\mathbf{x})$, given by

$$f(\mathbf{x}) = \sum_{s=1}^{g} \lambda_s \alpha_s(\mathbf{x}, \boldsymbol{\mu}_s \boldsymbol{\Sigma}_s). \tag{13.1}$$

In this expression $\mathbf{x}$ is a vector of order $p$ representing measurements for an individual on $p$ variates; $\lambda_s$, where

$$0 \leqslant \lambda_s \leqslant 1 \text{ and } \Sigma \lambda_s = 1,$$

is the proportion of individuals in the overall population drawn from the $s$th sub-population; while $\alpha_s(\mathbf{x}, \boldsymbol{\mu}_s, \boldsymbol{\Sigma}_s)$ is the multivariate normal density function

$$(2\pi)^{p/2} |\boldsymbol{\Sigma}_s|^{-1/2} \exp. -\tfrac{1}{2}(\mathbf{x} - \boldsymbol{\mu}_s)' \boldsymbol{\Sigma}_s^{-1} (\mathbf{x} - \boldsymbol{\mu}_s),$$

with mean vector $\boldsymbol{\mu}_s$ and covariance matrix $\boldsymbol{\Sigma}_s$ in the $s$th sub-population. The problem is to identify and describe the sub-populations or clusters given a sample drawn from the overall population. Maximum likelihoods techniques are used to estimate the parameters, $\lambda_s$, $\boldsymbol{\mu}_s$ and $\boldsymbol{\Sigma}_s$, and each individual is then associated with a cluster according to his maximum $P(s/\mathbf{x})$ value where

$$P(s/\mathbf{x}) = P \text{ (of an individual belonging to the } s\text{th cluster} \mid \text{his vector of scores } \mathbf{x})$$

$$= \frac{\lambda_s \alpha_s(\mathbf{x}_s, \boldsymbol{\mu}_s, \boldsymbol{\Sigma}_s)}{f(\mathbf{x})}. \tag{13.2}$$

Associated with this method is a likelihood ratio test for $g$, the number of clusters. The main problem with this technique is that it requires large samples to obtain accurate parameter estimates. Little is known too about its robustness when the assumption of multivariate normality of the variates is in doubt.

## Evaluating solutions

Though it is natural for an investigator to try to name and interpret any clusters he finds in a particular analysis, 'naming' on its own is likely to be considered gratuitous unless the clusters can be shown to be stable and to have some useful purpose outside themselves. One way of demonstrating such properties would be to show that the clusters had predictive value with respect to variables other than those used in defining them. For example, Entwistle and

Bennett (1974) used cluster analysis on data describing psychological and educational characteristics of students. The student 'types' found were shown to be associated with differing levels of academic performance implying that the 'types' have some useful predictive properties. In the same study a measure of replication was obtained by randomly dividing the sample into two and performing separate analyses on each. The clusters obtained from each analysis were very similar giving some indication of their 'stability'.

In addition to assessing the predictive ability of clusters various other methods might be used in evaluating them. For example, several different clustering techniques might be used on the same data. Were these to produce widely differing results then the existence of well defined clusters in the data would be in question. Again graphical techniques such as those described later in this chapter might be used alongside clustering techniques as an aid to the evaluation and interpretation of the latter. Indeed experience testifies to the danger of reporting a particular solution obtained by a particular clustering technique as the 'correct' solution without any attempt to validate its stability and usefulness.

*Example* 13.1

As an illustrative example of a clustering technique the method of mixtures described in the previous section was applied to the well-known 'iris leaf' data (Fisher, 1936). These data consist of measurements of sepal length and width, and petal length and width, for fifty plants of each of three species of iris, namely, *Iris setosa, Iris versicolor*, and *Iris virginica.* Although the correct classification into types of each of the 150 irises in the sample is known, the aim here is to try to discover the types without using this prior infor-

Table 13.1 Likelihood ratio for number of types in Fisher's iris leaf data

| *Numbers of type* | | $\chi^2_{28}$ | *Significance level* |
|---|---|---|---|
| *Null hypothesis* | *Alternative* | | |
| 1 | 2 | 320.0 | $10^{-10}$ |
| 2 | 3 | 66.0 | $10^{-4}$ |
| 3 | 4 | 43.0 | 0.03 |

mation. The analysis was performed using the NORMIX program (Wolfe, 1967), for fitting mixtures of multivariate normal distributions. This program was run on the data using hypotheses of *one* type, *two* types, *three* types and *four* types respectively, each

Table 13.2 Initial and final estimates of some parameters in the three-type solution

Type 1

| *Variate* | *Initial means* | *Final means* |
|---|---|---|
| S.L. | 5.0 | 5.0 |
| S.W. | 3.4 | 3.4 |
| P.L. | 1.5 | 1.5 |
| P.W. | 0.2 | 0.2 |

Initial Proportion = 0.333; Final proportion = 0.333;

Type 2

| | *Initial means* | *Final means* |
|---|---|---|
| S.L. | 6.0 | 5.9 |
| S.W. | 2.8 | 2.8 |
| P.L. | 4.5 | 4.2 |
| P.W. | 1.4 | 1.3 |

Initial proportion = 0.433; Final proportion = 0.299;

Type 3

| | *Initial means* | *Final means* |
|---|---|---|
| S.L. | 6.8 | 6.5 |
| S.W. | 3.1 | 2.9 |
| P.L. | 5.7 | 5.5 |
| P.W. | 2.1 | 2.0 |

Initial proportion = 0.233; Final proportion = 0.367;

Convergence took place after 21 iterations.

hypothesis being tested against the previous one by the likelihood ratio test already mentioned. In each case initial parameter estimates for the maximum likelihood equations were generated by the program itself using an hierarchial clustering method. But if 'good' estimates were known in advance, from the use of other methods or from prior information, these might usefully be employed instead.

Table 13.1 gives the results of the likelihood ratio tests for the different numbers of types and, testing at the 1% level, we are led to accept the three-type hypothesis in favour of the others.

Table 13.2 shows the initial and final estimates of type means for each of the four variables for the three-type solution together with the initial and final estimates of the proportions in each type.

Examination of the probabilities of group membership showed that the *Iris setosa* (Type 1) were completely recovered, as were *Iris virginica* (Type 3). However, five plants of *Iris versicolor* (Type 2) had higher probabilities of belonging to Type 3 and therefore these could be regarded as misclassified. From the results we see that the NORMIX program has essentially recovered the known structure amongst the 150 iris plants.

## VISUAL REPRESENTATIONS OF MULTIVARIATE DATA

In the description and statistical analysis of sets of data visual display and graphical presentation are invaluable and, as the mass media well know, are generally more helpful in communication of information than mere tables of numbers since, in general, people assimulate information more readily when it is presented in pictorial form. For univariate data histograms, dot diagrams, pie-charts and ideographs are commonly used and for bivariate data scatter-grams and trend lines are helpful. But methods for the visual representation of multivariate data are less common and less well known. Such data, consisting of a set of $p$ measurements or scores for each of $n$ individuals or objects, may be conceptualized in geometric terms as defining $p$ points in $n$-dimensional space, or as $n$ points in $p$-dimensional space. Useful though this is from a mathematical viewpoint it is limited from a visual viewpoint as we are restricted in viewing things to a 3-dimensional space. Indeed a common practice is to view the variates in subgroups of two or three, but this can be misleading as Cattell and Coulter (1966) have demonstrated.

In this section various methods for obtaining a visual representation of multivariate data will be described. Some of these operate directly on the $n \times p$ matrix of raw data; others require the $n \times n$ matrix of inter-individual similarities, dissimilarities or distances and are particularly relevant when the data actually arise in the form of a 'preference' matrix. In general all are best regarded as tools for a preliminary examination of multivariate data, which may give qualitative insight into their 'structure' and indicate the most appropriate models to use for more detailed analyses.

## (I) *Ordination techniques*

Ordination techniques permit the mapping of the $n$, $p$-dimensional vectors from $p$-space to a lower dimensional space, say $p^*$. If $p^* = 2$ or 3 this enables the data to be inspected visually and any patterns to be identified. A commonly used ordination technique is principal component analysis (Chapter 4). By plotting the data in the space of the first two or three components an indication of any structure present may be obtained. A more complex technique is that developed by Sammon (1969) called Non-Linear Mapping or NLM for short. This attempts to obtain a lower dimensional representation of the data by minimizing a function of the differences in the distances, $d_{ij}$, between points in the original $p$-space and those in the reduced $p^*$-space, denoted by $d^*_{ij}$. The function is simply a weighted sum of squares of these differences and is

$$E = \frac{1}{\sum_{i<j} d_{ij}} \sum_{i<j} \frac{(d_{ij} - d^*_{ij})^2}{d_{ij}}.$$

This particular weighting tends to preserve 'local' structure in the sense that each point in $p^*$-space bears approximately the same relationship to its near neighbours as it did in $p$-space, whilst its relationship to points relatively distant in $p$-space may be quite different in $p^*$-space. Starting with an arbitrary two-dimensional configuration (assuming we are interested in the $p^* = 2$ solution) points are adjusted using a steepest descent procedure so that $E$ is minimized. This gives a final two dimensional configuration which retains much of the structure present in the original $p$-dimensions.

As noted above some ordination techniques operate directly on the matrix of inter-individual similarities, dissimilarities or distances. Probably the simplest of these is the method described by Gower

(1966) and known as *principal co-ordinates analysis.* This is a latent roots and vectors method which recovers co-ordinates for points from a matrix giving distances between them. In many cases the first two or three of these co-ordinates may give a reasonable representation of the distances; consequently a plot can be made and a visual search for 'structure' undertaken.

Another method, particularly suitable when the elements of a similarity or dissimilarity matrix can only be assumed to have ordinal significance, is that suggested by Shepard (1962), and Kruskal (1964a, 1964b). It is known as *non-metric multidimensional scaling* and endeavours to find a set of points in a $p^*$-dimensional space such that the distances between them in this space are monotonically related to their similarities or dissimilarities. For a given value of $p^*$ the monotonicity property cannot in general be completely satisfied, and some means is needed of assessing the extent to which a configuration falls short of the requirement. For the purpose, Kruskal introduces a measure called 'stress' and chooses the configuration that minimizes this measure; stress is given by

$$S = \frac{\Sigma(d_{ij} - \hat{d}_{ij})^2}{\Sigma d_{ij}^2},$$

where the $d_{ij}$'s are the distances between points in the $p^*$-space and the $\hat{d}_{ij}$ are a set of numbers which are monotonic with the similarities or dissimilarities; that is, if the measured dissimilarities, $s_{ij}$, are ordered from lowest to highest, then the corresponding $\hat{d}_{ij}$'s also range from lowest to highest. In other words by minimizing $S$, a set of points is obtained for which the inter-point distances (which may be Euclidean or some other metric distance), are as close as possible to a set of numbers known to be monotonic with the measured dissimilarities. Only *ordinal* significance of the similarities or dissimilarities is implied in this method. The actual minimization technique is described in detail in Kruskal (1964b), but essentially it begins with an arbitrary $p^*$-configuration and proceeds in a stepwise manner making successive adjustment to the co-ordinates so as to decrease the stress. The problem of sub-optimal solutions arises here as it does with some clustering techniques, and the problem of deciding on the most suitable value of $p^*$ parallels that of choosing a value for $g$ in clustering. Of course, if a visual representation is the prime requirement, $p^*$ is restricted to the values 2 or 3.

The usefulness of ordination techniques is judged by how well the inherent structure of the data is preserved under the mapping from the original $p$-dimensional space to the lower $p^*$-dimensional space, or, in other words, by how well the original distances between individuals are maintained in the $p^*$-space. Various methods of measuring 'goodness of fit' have been proposed and are discussed by Cormack (1971) and by Gower (1970, 1971).

### (II) *Andrews' method for obtaining plots of multivariate data*

Andrews (1972) describes a method of function plotting which also enables one to obtain a visual representation of multivariate data; each $p$-dimensional point $\mathbf{X} = (x_1 x_2 \ldots x_p)$ defines a function

$$f_x(t) = x_1/\sqrt{2} + x_2 \sin t + x_3 \cos t + x_4 \sin 2t + x_5 \cos 2t + \ldots$$

This function is then plotted over the range—$\pi \leqslant t \leqslant \pi$. A set of $p$-dimensional points will now appear as a set of lines on the plot. Since this particular function preserves distances, close points appear as lines which remain close together for all values of $t$, whilst distant points will be represented by lines which remain apart for at least some value of $t$. Examination of plots for bands of lines remaining together for all $t$, enables clusters of points in the original $p$-dimensional space to be identified. In general examination of the plotted functions may show evidence of 'clusters', 'outliers' and other features of the data.

### (III) *Using faces to represent multivariate data*

Chernoff (1973) describes a technique which represents each $p$-dimensional observation by the cartoon of a face, whose features, such as curvature of chin, shape of nose, etc., are determined by the values taken by particular variables. A sample of points in $p$-dimensional space is thus represented by a collection of faces and 'clusters' may be identified by grouping similar faces.

A parallel technique is suggested by Frith (1974) who represents an observation by a face which varies between two extremes depend-

Table 13.3 Variable features in cartoon face

| | | |
|---|---|---|
| Upper hair, | Chin, | Lower hair, |
| Eye size, | Mouth size, | Eye space, |
| Eye slant, | Mouth curve, | Face size. |

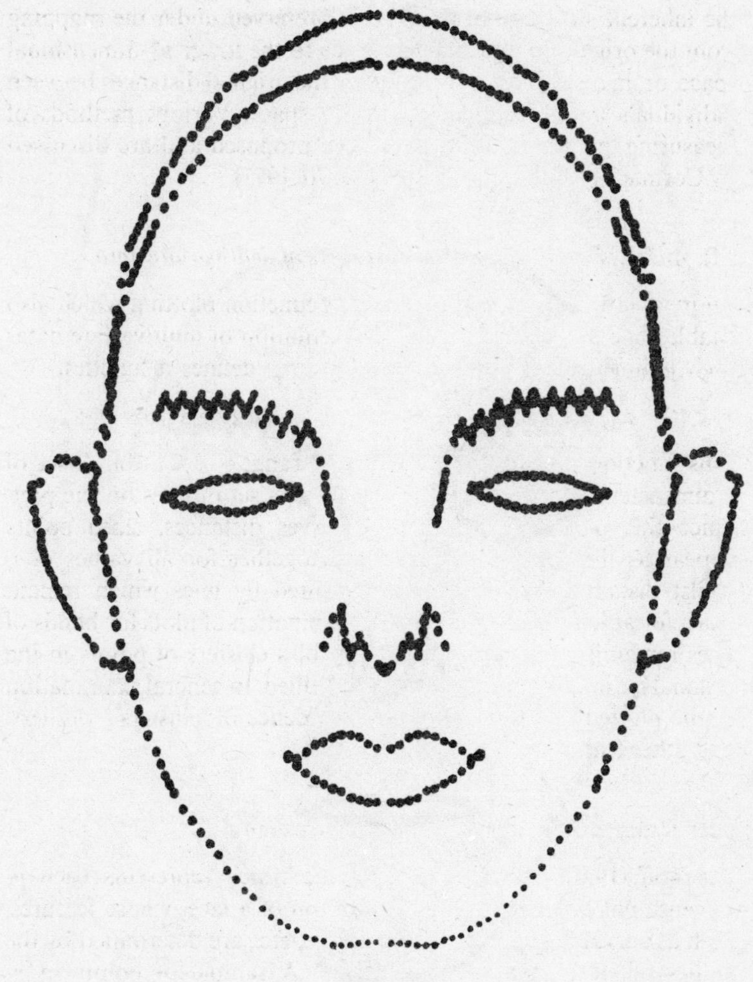

Fig. 13.1.

ing on the variable values. In this method the features used to represent variables are shown in Table 13.1, and the two extreme faces are shown in Figures 13.1 and 13.2. Up to nine variables can be represented, but the number could be increased by adding other features.

Of course the observations could just as easily be represented by features of other familiar objects, but the representation

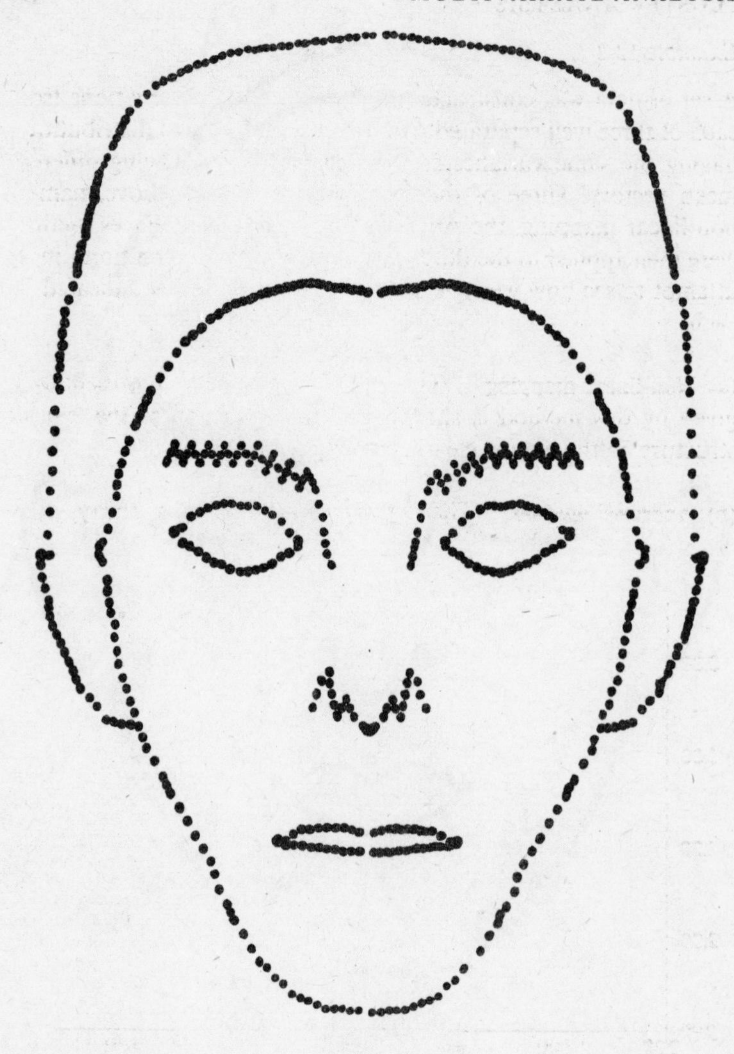

Fig 13.2.

in terms of faces may be the more useful since people are accustomed to studying and reacting to faces and to classifying them into ethnic types. In consequence, they may readily be able to classify in a useful way a set of things whose characteristics are expressed in terms of facial appearance (see Everitt and Nicholls, 1975).

*Example* 13.2

A set of data was constructed by sampling ten observations from each of three well separated five-dimensional normal distributions, having the same variance-covariance matrix but having different mean vectors. Three of the techniques described above, namely non-linear mapping, the Andrews' method and the Faces method were then applied to the thirty-five-dimensional observations, in an attempt to see how well the structure in the data was indicated by each.

**(a) Non-linear mapping.** The two-dimensional representation given by this method is shown in Figure 13.3, where the 'cluster structure' in the data is clearly evident.

**(b) Andrews' method.** The function plots of the thirty, five-

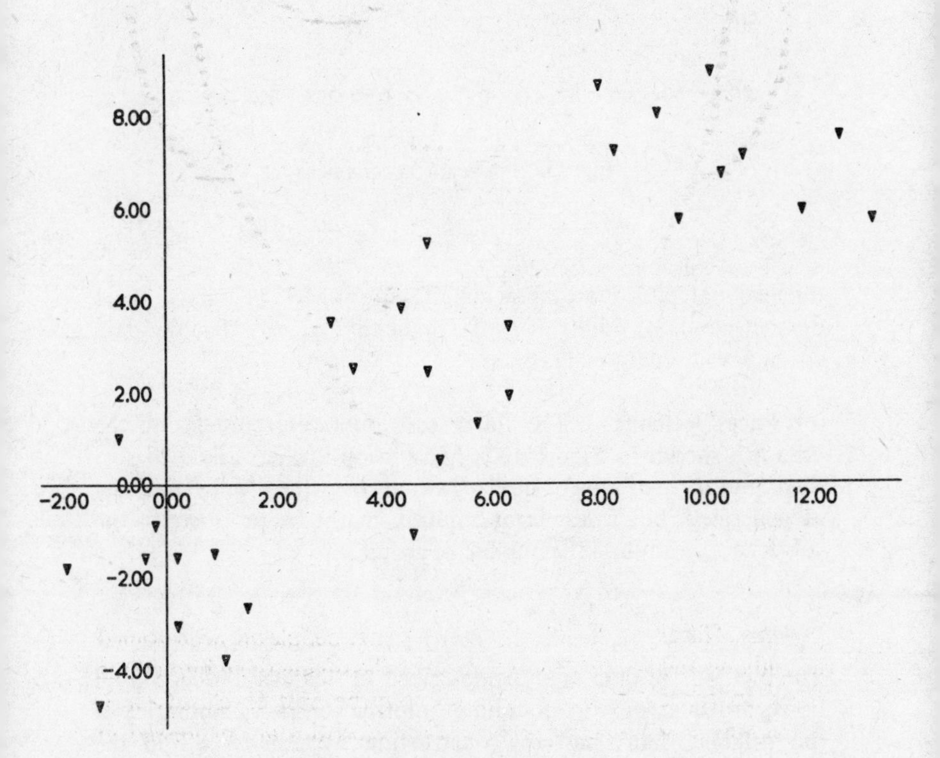


Fig. 13.3. *Non-linear mapping.*

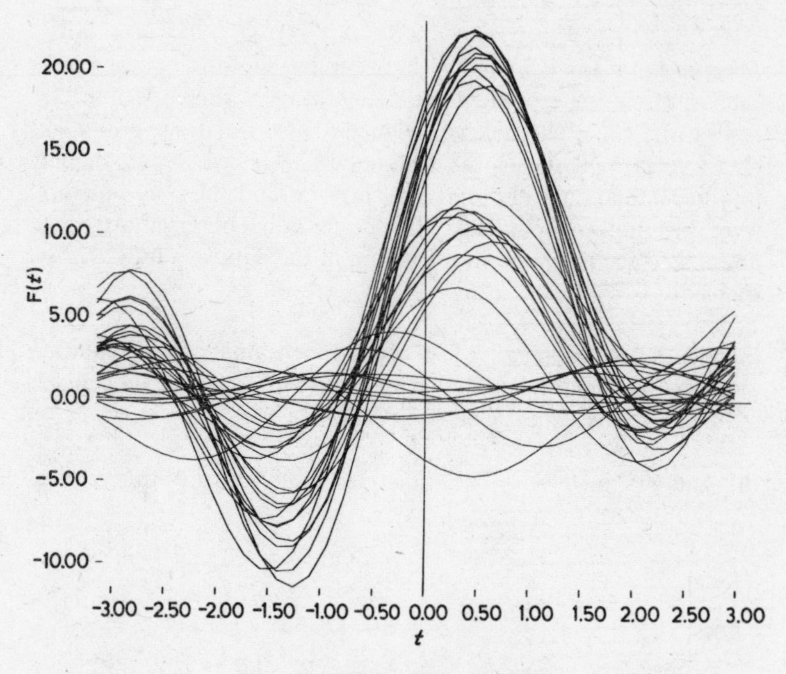


Fig. 13.4. *Andrews' function plots.*

dimensional points are shown in Figure 13.4. They show three distinct bands of points at $t = 0.5$ indicating clearly that the data contain well separated clusters.

**(c) 'Faces' technique.** The thirty cartoon faces representing the data are shown in Figure 13.5. Most people when asked to sort them into *three* groups of similar faces recover the groups of points as generated, but a less clear solution might be produced if the subjects were not told the number of groups.

## Summary

The methods of data analysis described in this chapter should all be regarded as aids in obtaining 'information-rich summaries of the original data', and in 'constructing hypotheses within the subject involved' (quotations from Sibson, 1972). Like other methods

Fig. 13.5. *Cartoon faces.*

of multivariate analyses they are *not* a panacea for poor data, and the user should bear in mind any implicit assumptions (e.g. that clusters are *spherical*) involved. However, used in *association* with other multivariate techniques, such as principal component and canonical variate analyses, they generally lead to greater insight into complex patterns which may exist in multivariate data.

# References

Aitken, A.C. (1934) Note on selection from a multivariate normal population, *Proc. Edin. Math. Soc.*, **4**, 106–110.

Andrews, D.F. (1972) Plots of high dimensional data. *Biometrics*, **28**, 125–136.

Armitage, P. (1971) *Statistical Methods in Medical Research.* Blackwell: Oxford.

Bartlett, M.S. (1937) The statistical conception of mental factors, *Brit. J. Psychol.*, **28**, 97–104

Bartlett, M.S. (1947) The general canonical correlation distribution *Ann. Math. Stat.*, **18**, 1–17.

Bartlett, M.S. (1948) Internal and external factor analysis, *Brit. J. Psychol. (Stat. Section)*, **1**, 73–81.

Bartlett, M.S. (1950) Tests of significance in factor analysis, *Brit. J. Psychol. (Stat. Section)*, **3**, 77–85.

Bartlett, M.S. (1953) Factor analysis in psychology as a statistician sees it. In: *Essays on Probability and Statistics.* Methuen: London.

Birch, M.W. (1963) Maximum Likelihood in three-way contingency tables. *J. Roy. Stat. Soc. (B)*, **25**, 220–233.

Bishop, Y.M.M. (1969) Full contingency tables, logits and split contingency tables, *Biometrics*, **25**, 383–399.

Bock, R.D. (1975) *Multivariate Statistical Methods.* Wiley: New York.

Box, G.E.P. (1949) A general distribution theory for a class of likelihood criteria, *Biometrika*, **36**, 317–346.

Burt, C. (1909) Experimental tests of general intelligence, *Brit. J. Psychol.*, **3**, 94–177.

Burt, C. (1917) *The Distribution and Relations of Educational Abilities.* King: London.

Burt, C. (1949) The structure of the mind: a review of the results of factor analysis, *Brit. J. educ. Psychol.*, **19**, 100–199.

Burt, C. (1950) The evidence for the concept of intelligence, *Brit. J. educ. Psychol.*, **25**, 158–177.

Cattell, R.B. and Coutler, M.A. (1966) Principles of behavioural taxonomy and the mathematical basis of the taxonome computer program, *Brit. J. math. statist. Psychol.*, **19**, 237–269.

Chernoff, H. (1973) Using faces to represent points in $k$-dimensional space graphically, *J. Am. Statis. Assoc.*, **68**, 361–368.

Claringbold, P.J. (1958) Multivariate quantal analysis, *J. Roy. Stat. Soc., B.*, **20**, 398–405.

Cormack, R.M. (1971) A review of classification, *J. Roy. Stat. Soc. (A)*, **134**, 321–367.

Cochran, W.G. (1970) Some effects of errors of measurement on multiple correlation, *J. Am. Statist. Ass.*, **65**, 22–34.

Efroymson, M.A. (1967) Multiple regression analysis. In *Mathematical Methods for Digital Computers*, eds. A. Ralston, and H.S. Wilf, Wiley: New York.

Emmett, W.G. (1942) *An Inquiry into the Prediction of secondary-School Success*. Univ. London Press: London.

Entwistle, N.J. and Bennett, S.N. (1974) Cluster analysis and educational research. Abstract in *Mult. Beh. Research*, **9**, 498.

Everitt, B.S. (1974) *Cluster Analysis*. Heinemann: London.

Everitt, B.S. (1977) *The Analysis of Contingency Tables*. Chapman and Hall: London.

Everitt, B.S. and Nicholls, P.G. (1975) Visual techniques for representing multivariate data, *The Statistician*, **1**, 24, 37–49.

Finn, J.D. (1974) *A General Model for Multivariate Analysis*. Holt-Reinhart and Winston: New York.

Fisher, R.A. (1936) The use of multiple measurements in taxonomic problems, *Annals of Eugenics*, **7**, 179–188.

Fleishman, E.A. and Hempel, W.E. (1954) Changes in factor structure of a complex psychomotor test as a function of practice. *Psychometrika*, **19**, 239–252.

Frith, C. (1974) Personal Communication.

Galton, F. (1869) *Hereditary Genius*. Horizon Press: New York. Paperback edition 1952.

Goodman, L.A. (1971) The analysis of multidimensional contingency tables, *Technometrics*, **13**, 33–55.

Gower, J.C. (1966) Some distance properties of latent root and vector methods used in multivariate analysis, *Biometrika*, **53**, 325–338.

Gower, J.C. (1970) Classification and geology. *Rev. ISI.*, **38**, 35–41.

Gower, J.C. (1971) Statistical methods of comparing different multivariate analyses of the same data. *Mathematics in the Archaeological and Historical Sciences*. eds. F.R. Hodson, D.G. Kendall and P.A. Tautu; University Press: Edinburgh.

Guilford, J.P. (1967) *The Nature of Human Intelligence*. McGraw-Hill: New York.

Healy, M.J.R. (1965) Descriptive uses of discriminant functions. In *Mathmatics and Computer Science in Biology and Medicine*. H.M. Stationery Office: London.

Hendrickson, A.E. and White, P.O. (1964) Promax: a quick method for rotation to oblique simple structure, *Brit. J. Statis. Psychol.* **17**, 65–70.

Hodges, S.D. and Moore, P.G. (1972) Data uncertainties and least squares regression, *J. App. Statist.*, **21**, 185–195.

Horst, P. (1961) Generalised canonical correlations and their application to experimental data, *J. Clin. Psychol.* (monog. supp.), No. 14, 331–347.

Hotelling, H. (1933) Analysis of a complex of statistical variables into principal components, *J. educ. Psychol.*, **24**, 417–441, 498–520.

Hotelling, H. (1935) The most predictable criterion, *J. educ. Psychol.*, **26**, 139–142.

Hotelling, H. (1936) Relations between two sets of variates, *Biometrika*, **28**, 321–377.

Jardine, N. and Sibson, R. (1968) The construction of hierarchic and non-hierarchic classifications, *Comp. J.*, **11**, 117–184.

Jöreskog, K.G. (1966) UMFLA—A computer program for unrestricted maximum likelihood factor analysis. *Research Memorandum* 66–20, Princeton N.J., Educational Testing Service.

Jöreskog, K.G. and Gruvaeus, G. (1967) RMLFA—A computer program for restricted maximum likelihood factor analysis, *Research Memorandum* 67–71, Princeton, N.J., Educational Testing Service.

Jöreskog, K.G. (1967) Some contributions to maximum likelihood factor analysis, *Psychometrika*, **32**, 443–482.

Kelley, T.L. (1928) *Crossroads in the Mind of Man*. Stanford Univ. Press: Stanford, California.

Kendall, M.G. and Lawley, D.N. (1956) The Principles of factor analysis. *J. Roy. Stat. Soc. (A)*, **119**, 83–84.

Kruskal, J.B. (1964a) Multidimensional scaling by optimising goodness of fit to non-metric hypothesis, *Psychometrika*, **29**, 1–27.

Kruskal, J.B. (1964b) Non-metric multidimensional scaling: a numerical method, *Psychometrika*, **29**, 115–129.

Lawley, D.N. (1943) A note on Karl Pearson's selection formulae, *Proc. Roy. Soc. Edin.*, **62**, 28–30.

Lawley, D.N. (1944) The factorial analysis of multiple item tests, *Proc. Roy. Soc. Edin*, **62**, 74–82.

Lawley, D.N. and Maxwell, A.E. (1971) *Factor Analysis as a Statistical Method*. Butterworths: London.

Lawley, D.N. and Maxwell, A.E. (1973) Regression and factor analysis, *Biometrika*, **60**, 331–338.

Ling, R.F. (1972) On the theory and construction of *k*-clusters, *Comp. J.*, **15**, 326–332.

Lord, F.M. and Novick, M.R. (1968) *Statistical Theories of Mental Test Scores*. Addison-Wesley: Reading, Mass.

Lubin, A. (1950) Linear and non-linear discriminating functions, *Brit. J. math. and statist. Psychol.*, **3**, 90–103.

Lubin, A. (1957) Some formulae for use with suppressor variables, *Educ. Psychol. Meas.*, **17**, 286–296.

Macdonell, W.R. (1902) On criminal anthropometry and the identification of criminals, *Biometrika*, **1**, 177–227.

Maxwell, A.E. (1961) Canonical variate analysis when the variables are dichotomous, *Educ. Psychol. Meas.*, **21**, 259–271.

Maxwell, A.E. (1964) *Analysing Qualitative Data*, Methuen: London.

Maxwell, A.E. (1972a) Factor analysis: Thomson's sampling theory recalled, *Brit. J. math. statist. Psychol.*, **25**, 1–21.

Maxwell, A.E. (1972b) The WPPSI: a marked discrepancy in the correlations of the subtests for good and poor readers, *Brit. J. math. statist. Psychol.*, **25**, 283–291.

Maxwell, A.E., Fenwick, P.B.C., Fenton, G.W. and Dollimore, J. (1974) Reading ability and brain function: a simple statistical model. *Psychol. Medicine*, **4**, 274–280.

Meredith, W. (1964) Canonical correlations with fallible data, *Psychometrika*, **29**, 55–66.

Morrison, D.F. (1967) *Multivariate Statistical Methods*, McGraw-Hill: New York.

Moursy, E.M. (1952) The hierarchical organization of cognitive levels, *Brit. J. Psychol. stat. sect.*, **5**, 151–180.

Nelder, J.A. (1965) The analysis of randomized experiments with orthogonal block structure, *Proc. Roy. Soc. London, A*, **283**, 147–162, 163–178.

Pearson, K. (1901) On lines and planes of closest fit to a system of points in space, *Phil. Mag.*, **2**, 6th series, 557–572.

Pearson, K. (1904) On the laws of inheritance in man, *Biometrica*, **3**, 131–190.

Pearson, K. (1920) Notes on the history of correlation, *Biometrika*, **13**, 15–34.

Radhakrishna, S. (1964) Discrimination analysis in medicine, *Statistician*, **14**, 147–167.

Rao, C.R. (1948) The utilization of multiple measurements in problems of biological classification, *J. Roy. Stat. Soc., B*, **10**, 159–193.

Sammon, J.W. (1969) A non-linear mapping for data structure analysis, *IEEE Trans. Computers C* **18**, 401–409.

Searle, S.R. (1966) *Matrix Algebra for the Biological Sciences.* Wiley: New York.

Shepard, R.N. (1962) The analysis of proximities: multi-dimensional scaling with an unknown distance function, *Psychometrika*, **27**, 125–139; 219–246.

Simon, F.H. (1971) *Prediction Methods in Criminology*, H.M. Stationery Office: London.

Smith, C.A.B. (1946) Some Examples of Discrimination. *Annals of Eugenics* XIII 2272–383.

Sneath, P.H.A. and Sokal, R.R. (1963) *Principles of Numerical Taxonomy.* Freeman: London.

Spearman, C. (1904) The proof and measurement of association between two things, *American J. Psychol.* **15**, 88–103.

Spearman, C. (1904) 'General intelligence', objectively determined and measured, *Amer. J. Psychol.* **15**, 201–293.

Thomson, G.H. (1919) On the cause of hierarchical order among correlation coefficients, *Proc. Roy. Soc., A*, **95**, 400–408.

Thomson, G.H. (1939) *The Factorial Analysis of Human Ability*. Univ. London Press: London.

Thurstone, L.L. (1938) *Primary Mental Abilities*, Psychometric Monog.

Welch, B.L. (1939) Note on discriminant functions, *Biometrika*, **31**, 218–220.

Williams, E.J. (1952) Use of scores for the analysis of association in contingency tables, *Biometrika*, **39**, 274–289.

Williams, W.T., Lance, G.N., Dale, M.B. and Clifford, H.T. (1971) Controversy concerning the criteria for taxonometric strategies, *Comp. J.* **14**, 162–165.

Wolfe, J.H. (1967) NORMIX; computational methods for estimating the parameters of multivariate normal mixtures of distributions. *Research Memorandum*, SRM 68–2, U.S. Naval Personnel Research Activity, San Diego.

Wolfe, J.H. (1969) Pattern clustering by multivariate mixture analysis. *Research Memorandum*, SRM 69–17, U.S. Naval Personnel Research Activity, San Diego.

Wolfe, J.H. (1970) Pattern clustering by multivariate mixture analysis. *Multiv. Behav. Res.*, **5**, 329–350.

Yule, W., Berger, M., Butler, S., Newham, V. and Tizard, J. (1969) The WPPSI: an empirical evaluation with a British sample. *Brit. J. educ. Psychol.*, **39**, 1–13.

# Index

# Index

Agglomerative hierarchical techniques, 138
Aitken, A.C., 75, 153
Analysis of covariance, 119
Analysis of variance, 115–128
  generalization, 126
  linear model, 115
  multivariate, 129–135
  tests, 133
  two-way, 115
Andrews, D.F., 146, 149, 153
  function plotting method, 146, 149
Armitage, P. 96, 153
Assignment techniques, 136
Attenuation, correction for, 6, 80

Bartlett, M.S., 5, 12, 49, 58, 67, 88, 93, 153
Bennett, S.N., 141, 153
Berger, M., 157
Birch, M.W., 114, 153
Bishop, Y.M.M., 114, 153
Bivariate normality, 111
Box, G.E.P., 105, 153
Burt, C., 5, 10, 11, 153

Canonical correlations, 85–93
Canonical variate analysis, 94, 96
Canonical variates, scaling, 98, 99
Cattell, R.B., 143, 153
Characteristic root, 30
Chernoff, H., 146, 153
Claringbold, P.J., 96, 154
Classifying individuals, 44
Clifford, H.T., 157
Clustering, single linkage, 138
Cluster analysis, 136
Cochran, W.G., 81, 154
Coefficients, partial regression, 70, 72
  similarity, 137
Communality, 9, 47, 48
Contingency tables, analysis of, 106–114
  multidimensional, 114
Continuum, 5, 18
Cormack, R.M., 137, 146, 154
Correlation, partial, 8
Coulter, M.A., 143, 153
Covariance, estimate of, 15
Criterion, clustering, 139
  external, 2
  most predictable, 93

Dale, M.B., 157
Designs, non-orthogonal, 123
Determinants, 25
Determinantal equation, 31
Determinants, 25
Difficulty component (factor), 45
Dimensions, 18, 47
Discriminant function, 94, 136
  quadratic, 102
Discrimination, sensory, 10
Distance measures, 137

Efroymson, M.A., 79, 154
Eigenvalue, 30
Emmett, W.G., 76, 78, 80, 154

Entwistle, N.J., 140, 154
Everitt, B.S., 42, 137, 138, 148, 154
Experimental designs, 115

Factor analysis, 5, 46–82
  and regression, 81
  confirmatory, 60
  exploratory, 60
  external, 93
  restricted model, 60
Factor, general, 10, 11, 12, 56, 63
  group, 10, 12, 63
  loadings, 47, 48
  model, 15, 47, 59
  pattern, 60, 63, 64
  rotation, 53
  scores, 66–68
  specific, 12
  spurious, 20
Factors, correlated, 49, 61, 69
  errors of prediction, 68
  oblique, 68
  orthogonal, 60
Fisher, R.A., 5, 96, 141, 154
Fleishman, E.A., 55, 154
Frith, C., 146, 154

Galton, F., 1, 2, 3, 4, 5, 6, 8, 18, 154
Goodman, L.A., 114, 154
Gower, J.C., 45, 144, 146, 154
Gruvaeus, G., 61, 155
Guilford, P.E., 11

Healy, M.J.R., 96, 154
Hempel, W.E., 55, 154
Hendrickson, A., 54, 154
Hierarchical correlation matrices, 8
Hill-climbing algorithm, 139
Hodges, S.D., 81, 154
Horst, P., 86, 154
Hotelling, H., 4, 85, 93, 155

Intelligence, 5, 6, 10, 11
Item analysis, 17

Jardine, N., 138, 155
Jöreskog, K.G., 49, 57, 61, 155

Kelley, T.L., 10, 155
Kendall, M.G., 58, 155
Kronecker product, 126
Kruskal, J.B., 145, 155

Lagrange multiplier, 36
Lance, G.N., 157
Lawley, D.N., 20, 35, 37, 46, 49, 58, 66, 81, 82
Likelihood ratio test, 140, 143
Linear independence, 16
Ling, R.F., 137, 155
Lord, F.M., 80, 155
Lubin, R.D., 78, 103, 155

Macdonell, W.R., 3, 4, 42, 155
MANOVA, 129–135
Mapping, non-linear, 149
Matrices, addition of, 22
  basis, 126
  comparing covariance, 105
  direct product of, 126
  inverse, 26
  Kronecker product of, 126
  orthogonal, 33
  rectangular, 23
  triangular, 14
Matrix, correlation, test of significance, 49
  covariance, 39, 46
  design, 117
  diagonal, 23
  dispersion (*see* covariance matrix)
  elements of, 22
  latent roots of, 30
  latent vectors of, 30
  null or zero, 23
  of loadings, 48
  order of, 44
  positive definite, 35
  rank of, 29

rotation, 34
singular, 29
square, 22
symmetric, 22, 37
trace of, 30
transpose of, 23
triangular, 34, 37
unit (identity), 23
Maxwell, A.E., 11, 12, 35, 37, 46, 57, 58, 61, 65, 66, 81, 82, 96, 106, 111
Measurement, errors of, 6, 14, 15
Meredith, W., 86, 154
Metric, 19
inequality, 137
Moore, P.G., 81, 154
Morrison, D.F., 41, 46, 100, 103, 134, 156
Moursy, E.M., 11, 156
Multiple correlation coefficient, 2, 70, 73, 74, 82, 84
Multiple linear regression, 70–84
Multivariate data, faces to represent, 146
Multivariate normal distribution, 20, 46, 140
distributions, mixtures of, 139

Nelder, J.A., 128, 156
Nicholls, P.G., 148, 154
Normix program, 142, 143
Novick, M.R., 80, 155

Oblique factors, 68

Parameters, constraints on, 116
estimates of, 14
fixed, 60, 61
free, 60, 61
population, 13
Pearson, K., 1, 2, 4, 75, 77, 156
Phi-coefficients, 20, 45, 111
Principal co-ordinate analysis, 145
Principal component analysis, 2, 39–45, 58,
Promax method, 54

Q-technique, 44
Quadratic forms, 35, 103

Radhakrishna, S., 96, 156
Rao, C.R., 96, 156
Regression and Factor Analysis, 81–84
Regression coefficient, 1, 2, 81
Reliability, 6, 13
Rotation of axes, 43

Sammon, J.W., 144, 156
Sampling, selective, 74
Scaling, non-metric multidimensional, 145
Shepard, R.N., 145, 156
Sibson, R., 138, 150, 155
Similarity measures, 137
Simon, F., 80, 156
Simultaneous equations, solution of, 27–29
Smith, C.A.B., 102, 156
Sneath, P.H.A., 137, 156
Sokal, R.R., 137, 156
Spearman, C., 6, 7, 8, 9, 10, 13, 17, 80, 156
Spearman's two-factor theory, 6, 10
Spencer, H., 5
Stress, measure of, 145

Technique, faces, 146
Techniques, graphical, 141
optimization, 146
ordination, 144
Terminology, 18
Tetrachoric correlation coefficient, 20
Tetrad difference, 9
Thomson, G. H., 11, 12, 18, 65, 66, 156
Thurstone, L.L., 10, 11, 156
Transformation, monotonic, 138

Variance, error, 50
residual, 52, 81, 82

specific, 9
zero estimate of residual, 57
Variate, criterion, 79
dependent, 70
dummy, 95
independent, 70
Variates, distribution of, 20
hypothetical, 39
linear constraints on, 16
linear independent, 16
non-random, 70
predictor, 70
residual, 50
selecting independent, 78
Varimax method, 54
Vectors, normalized, 32
product of two, 24
row and column, 23
Vernon, P.E., 10
Visual representation of data, 143

Welch, B.L., 96, 104, 157
White, P.O., 54, 154
Wilks, S.S., 5
Williams, E.J., 106, 110, 112, 157
Williams, W.T., 139, 157
Wishart, J., 5
Wolfe, J.H., 139, 142, 157

Yule, W., 61, 157